# Material Worlds:
# Intersections of Law, Science, Technology and Society

*Edited by*

Christopher Lawless and Alex Faulkner

WILEY-BLACKWELL

This edition first published 2012
Editorial organization © 2012 Cardiff University Law School
Chapters © 2012 by the chapter author

Blackwell Publishing was acquired by John Wiley & Sons in February 2007. Blackwell's publishing programme has been merged with Wiley's global Scientific, Technical, and Medical business to form Wiley-Blackwell.

*Editorial Offices*
350 Main Street, Malden, MA 02148-5020, USA
9600 Garsington Road, Oxford OX4 2DQ, UK

For details of our global editorial offices, for customer services, and for information about how to apply for permission to reuse the copyright material in this book please see our website at www.wiley.com/wiley-blackwell

*Registered Office*
John Wiley & Sons Ltd, The Atrium, Southern Gate, Chichester, West Sussex PO19 8SQ.

The right of Christopher Lawless and Alex Faulkner to be identified as the authors of the Editorial Material in this work has been asserted in accordance with the Copyright, Designs and Patents Act 1988.

All rights reserved. No part of this publication may be reproduced, stored in a retrieval system, or transmitted, in any form or by any means, electronic, mechanical, photocopying, recording or otherwise, except as permitted by the UK Copyright, Designs and Patents Act 1988, without the prior permission of the publisher.

Designations used by companies to distinguish their products are often claimed as trademarks. All brand names and product names used in this book are trade names, service marks, trademarks or registered trademarks of their respective owners. The publisher is not associated with any product or vendor mentioned in this book. This publication is designed to provide accurate and authoritative information in regard to the subject matter covered. It is sold on the understanding that the publisher is not engaged in rendering professional services. If professional advice or other expert assistance is required, the services of a competent professional should be sought.

*Library of Congress Cataloging-in-Publication Data*
Library of Congress Cataloging-in-Publication data is available for this book.

A catalogue record for this title is available from the British Library.

ISBN: 9781444361520

Set in the United Kingdom by Godiva Publishing Services Ltd
Printed in Singapore by Fabulous Printers Pte Ltd

# Contents

| | |
|---|---:|
| Introduction: Material Worlds: Intersections of Law, Science, Technology, and Society ......... *Alex Faulkner, Bettina Lange, and Christopher Lawless* | 1 |
| The Pragmatic Sanction of Materials: Notes for an Ethnography of Legal Substances .................. *Javier Lezaun* | 20 |
| The Regulation of Nicotine in the United Kingdom: How Nicotine Gum Came to Be a Medicine, but Not a Drug .... *Catriona Rooke, Emilie Cloatre, and Robert Dingwall* | 39 |
| The Donor-conceived Child's 'Right to Personal Identity': The Public Debate on Donor Anonymity in the United Kingdom ................................ *Ilke Turkmendag* | 58 |
| A Socio-legal Analysis of an Actor-world: The Case of Carbon Trading and the Clean Development Mechanism ...................... *Emilie Cloatre and Nick Wright* | 76 |
| Nanotechnology and the Products of Inherited Regulation............................................. *Elen Stokes* | 93 |
| The Emergence of Biobanks in the Legal Landscape: Towards a New Model of Governance ........... *Emmanuelle Rial-Sebbag and Anne Cambon-Thomsen* | 113 |
| The Legal Landscape for Advanced Therapies: Material and Institutional Implementation of European Union Rules in France and the United Kingdom ....... *Aurélie Mahalatchimy, Emmanuelle Rial-Sebbag, Virginie Tournay, and Alex Faulkner* | 131 |
| Bodies of Science and Law: Forensic DNA Profiling, Biological Bodies, and Biopower .................... *Victor Toom* | 150 |
| The Materiality of What? .............................. *Alain Pottage* | 167 |

# Introduction: Material Worlds: Intersections of Law, Science, Technology, and Society

ALEX FAULKNER,* BETTINA LANGE,** AND
CHRISTOPHER LAWLESS***

'*Material*':
n. The substance or substances out of which a thing is or can be made.
adj. Being both relevant and consequential; crucial; 'testimony material to the inquiry'.

## INTRODUCTION

The connections between law, science, and technology are ubiquitous and increasingly complex in contemporary societies. Symbols of these connections can be seen in road signs, food labels, safety certificates, passports, birth and death certificates, bank cards, medicinal prescriptions, property deeds, door security locks, motor vehicle performance tests, public street behaviour and its surveillance, and so on. Similar connections can also be found in a myriad of more hidden social spaces, such as biochemical laboratories, planning offices, industrial manufacturers, patent offices, crime

---

\* Department of Political Economy, King's College London, Strand, London WC2R 2LS, England
alex.faulkner@kcl.ac.uk
\*\* Centre for Socio-Legal Studies, University of Oxford, Manor Road, Oxford OX1 3 UQ, England
bettina.lange@csls.ox.ac.uk
\*\*\* Science Technology and Innovation, School of Social and Political Science, University of Edinburgh, Old Surgeons' Hall, High School Yards Edinburgh EH1 1LZ, Scotland
clawless@staffmail.ed.ac.uk

Early drafts of many of the articles in this volume were presented at a workshop held at King's College London on 5 April 2011. The editors thank Karen Yeung, Gearóid Ó Cuinn, and Asma Vranaki for their participation. For providing expert refereeing, we thank Marie Fox, Barbara Prainsack, Bettina Lange, Veerle Heyvaert, Sujatha Raman, Carole McCartney, Martyn Pickersgill, Michael Morrison, and Aaro Tupasela.

scenes, law courts, and parliaments. In all these cases, fundamental legally framed social relationships are at stake: contractual relationships, property ownership, as well as responsibilities, including liability, and rights arising from these. The connections between law, science, technology, and society continually evolve in interaction with each other, and to understand these connections represents a significant academic challenge.

In this volume, we focus on how developing interactions between social study of technology, science, and law can promote understanding of the ways in which legal processes and instruments interact with and produce 'material worlds'.[1] In doing so, we draw upon recent research that has explored the relationship between society and the material forms of science, technology, and nature. The contributions present challenging new perspectives on how legal practices relate to materiality, and invite theoretical reflection on how society even construes some material activities as 'law'.

The volume explores the practices which shape intersections of law, science, technology, and society, spanning spatial and temporal boundaries. The contributors explore how legal instruments and procedures co-emerge with new materialized realities. Legal landscapes evolve as local practices interact with globalization, leading to greater internationalization of governance and jurisdictions. In exploring this theme the collection draws together a wide variety of empirical topics, including the environment, medicine, forensic science, biotechnologies, nanotechnology, and human reproduction. The collection indicates new ways of understanding the continually evolving relationships between law, science, technology, and society. More specifically, it highlights developing synergies between scholarship in socio-legal studies (SLS) and science and technology studies (STS). To begin with, we will set out key developments in STS and SLS with reference to law-science-technology-society (LSTS) relationships.[2]

## SCIENCE AND TECHNOLOGY STUDIES

From the 1970s onwards, social studies of science and technology have grown in prominence and scope. STS is now a space where a variety of disciplinary perspectives engage with science and technology, including sociology, philosophy, anthropology, economics, and political science. Many key

---

1 We use the term 'material *worlds*' to emphasize the pluralities in the convergence of law, technology, and society. We note that other authors have used a similar term, for example, T. Pinch and R. Swedberg, *Living in a material world* (2008), (referring to a song by Madonna). We also have an eye to 'social worlds' theory (for example, A.E. Clarke, 'Social Worlds/Arenas Theory as Organizational Theory' in *Social Organization and Social Process: Essays in Honor of Anselm Strauss*, ed. D. Maines (1991) 119–58.
2 We use 'LSTS' here simply as shorthand to refer to relationships between law, science, technology, and society, rather than to define a specific field of study.

scholars in the field have made significant contributions to the social sciences beyond STS. The field now encompasses different, sometimes conflicting orientations, ranging from the originally largely Edinburgh-based 'Strong Programme' of the Sociology of Scientific Knowledge, in which the content of science itself is seen as shaped by social forces,[3] to the Social Construction of Technology (SCOT) programme,[4] which shows the social contingency in the design and use of material artefacts. Also other, more radical methodological perspectives have emerged such as Actor-Network Theory (ANT)[5] which questions the assumed division between 'social' and 'material' worlds.

A range of STS research has explored the relationships between law and science, highlighting the myriad complexities between what has been described as 'the two institutions, that, perhaps more than any other, are responsible for making order, and guarding against disorder, in contemporary societies'.[6] The regulation or, more broadly, 'governance' of technological risk has also attracted particular interest. Work here has focused on the demands administrators and policy makers place on scientists to deliver certainties in producing knowledge of regulatory relevance, and on the conduct of ethically acceptable scientific work. The boundaries between scientific and political concerns have been shown to be blurred and contested.[7] Studies of regulatory knowledge making have also shown that certain knowledge of a risk is often blocked by intractable ontological barriers. This, and the interplay of technical, governmental, and commercial factors means that 'regulatory science' (scientific knowledge created with a regulatory purpose) displays a highly contingent character.[8]

As well as scrutinizing this epistemology of regulation, STS has also addressed how regulation impacts upon the products of scientific endeavours and emerging technologies, particularly in the medical sector and increasingly in others such as energy and the environment. Important here is a focus on social and economic forces that account for scientific and technological innovation. Work on 'innovation systems'[9] is concerned with the shaping of

---

3 D. Bloor, *Knowledge and Social Imagery* (1976).
4 T. Pinch and W.E. Bijker, 'The Social Construction of Facts and Artefacts: Or How the Sociology of Science and the Sociology of Technology Might Benefit Each Other' (1984) 14 *Social Studies of Sci.* 399–441.
5 For an early example, see M. Callon, 'Some Elements of a Sociology of Translation: Domestication of the Scallops and the Fishermen of St Brieuc Bay' in *Power, Action and Belief: A New Sociology of Knowledge?*, ed. J. Law (1986).
6 S. Jasanoff, 'Making Order: Law and Science in Action' in *The Handbook of Science and Technology Studies*, eds. E.J. Hackett et al. (2007, 3rd edn.) 761.
7 S. Jasanoff, 'Contested boundaries in policy-relevant science' (1987) 17 *Social Studies of Sci.* 195–230.
8 A. Irwin et al., 'Regulatory science – towards a sociological framework' (1997) 29 *Futures* 17–31.
9 C. Freeman, 'Formal Scientific and Technical Institutions in the National System of Innovation' (1992) in *National Systems of Innovation: Towards a Theory of Innovation and Interactive Learning*, ed. B.-Å. Lundvall (1992) 169–87.

technologies and the human-made environment broadly, with key approaches framed in terms of sectoral systems[10] or 'innovation pathways'. There is increasing recognition of the importance of 'institutions' acting in a legal rather than economic sphere, shaping the direction of 'technological regimes'[11] or 'technological zones'.[12]

Social principles, religious values, and ethics are inscribed into such regimes or zones partly via law and regulation, and these are also now a major concern of STS, encapsulated in the so-called ELSI (ethical, legal, and social implications) approach. STS has also become interested in the design of fora in which social actors become *engaged with* law making. Most notable here is an ever-growing body of work, informed largely by democratic politics and a concern with human rights, expertise, and citizenship, that analyses the presence of or need for 'participatory governance' across legislative domains and techno-scientific fields.[13]

A key point of debate is whether the relation between regulation and innovation is one of 'regulatory lag' leaving the law to play catch-up. While some suggest that this is the case,[14] the possibility of anticipatory regulation suggests that this is not necessarily so. Legal regulation can also shape in advance the 'rules of the game' of material technology and the scientific process.[15] The *institutional* history, location, and local reasoning practices of law making and implementation have attracted attention. Latour's analysis of the Conseil d'Etat (which advises the French government on administrative law issues and is the highest court to adjudicate upon administrative law matters) is a touchstone here,[16] showing how the mundane admini-

---

10 For example, see F. Malerba, 'Sectoral systems of innovation: basic concepts' in *Sectoral Systems of Innovation: Concepts, Issues and Analyses of Six Major Sectors in Europe*, ed. F. Malerba (2004) 9–41.
11 For example, see J. Markard and B. Truffer, 'Technological innovation systems and the multi-level perspective: Towards an integrated framework' (2008) 32 *Research Policy* 596–615.
12 See A. Barry, 'Technological Zones' (2006) 9 *European J. of Social Theory* 239–53; A. Faulkner, 'Regulatory policy as innovation: constructing rules of engagement of a technological zone for tissue engineering in the European Union' (2009) 38 *Research Policy* 637–46.
13 For example, see R. Tutton et al., '*Myriad Stories: Constructing Expertise and Citizenship in Discussions of the New Genetics*' in *Science and Citizenship in a Global Context*, eds. M. Leach and I. Scones (2004).
14 B. Moses, 'Recurring Dilemmas: The Law's Race to Keep Up with Technological Change' (2007), UNSW Law Research Paper no. 2007-21, at <http://ssrn.com/abstract=979861>.
15 For example, see R. Bud, 'In the engine of industry: regulators of biotechnology, 1970–86' in *Resistance to new technology: nuclear power, information technology and biotechnology*, ed. M. Bauer (1999) 293–310; J. Chataway, J. Tait, and D. Wield, 'The governance of agro- and pharmaceutical biotechnology innovation: public policy and industrial strategy' (2006) 18(2) *Technology Analysis and Strategic Management* 1–17; Faulkner, op. cit., n. 12.
16 B. Latour, *The Making of Law: An Ethnography of the Conseil d'Etat* (2009).

strative materials of legal case files are implicated in the construction of legal precedent, shaping the terms in which tort law is produced and thus its adjudicative outcomes.

While these analyses of the constitutive effects of law are now emerging, the adversarial traditions of Anglo-American law have for some time constituted a fruitful site of STS work. Studies of judicial cross-examination have revealed how supposedly objective scientific knowledge is actually highly vulnerable. Legal professionals themselves have been shown to be capable of exposing the heterogeneous nature of scientific knowledge.[17] Moreover, the status of the 'expert' emerges as dependent on legal procedures as much as scientific evaluations. Courtrooms are spaces in which 'expert' status is performed and enacted, with counsel and judges playing as much of a role as witnesses.[18] Hence, courts have been revealed to also direct scientific and technological innovation, and thus shape society's relationships with new material objects.[19]

The adversarial courtroom has stimulated particular interest in forensic science. Forensic DNA profiling, and its implications for legal procedure and for the relationship between governments, police, and publics, has been a growing focus. In exploring how DNA databases have led to changes in the legislation concerning police rights to sample 'suspect' individuals, STS has highlighted the mutual interdependence of biotechnology and criminal law.[20] Other work has focused on enduring forms of forensic evidence such as fingerprint analysis and digital facial mapping.[21] Recent studies have shown how expert testimony can be regarded as an organizational and bureaucratic achievement as much as a technological one.[22]

Thus a range of strands of STS highlight the importance of law in the LSTS nexus, and a similar range of concerns with 'materiality'. Despite considerable attention to regulatory and governance matters, however, STS, paradoxically, has not paid much attention to 'black-letter' law. But attention to law, not just in its doctrinal manifestations but more widely in its social,

---

17 M. Lynch and S. Jasanoff, 'Contested Identities: Science, Law and Forensic Practice' (1998) 28 *Social Studies of Sci.* 675–86.
18 M. Lynch, 'Circumscribing Expertise: Membership Categories in Courtroom Testimony' in *States of Knowledge: The co-production of science and social order*, ed. S. Jasanoff (2004) 161–80.
19 S. Jasanoff, *Science at the Bar: Law, Science and Technology in America* (1995).
20 M. Lynch and R. McNally, 'Forensic DNA Databases and biolegality: The co-production of law, surveillance technology and suspect bodies' in *Handbook of Genetics and Society: Mapping the new genomic era*, eds. P. Atkinson, P. Glasner, and M. Lock (2009) 283–301.
21 S.A. Cole, *Suspect Identities: A History of Fingerprinting and Criminal Identification* (2001); G. Edmond et al., 'Law's Looking Glass: Expert Identification Evidence Derived From Photographic and Video Images'(2009) 20 *Current Issues in Criminal Justice* 337–77.
22 M. Lynch et al., *Truth Machine: The Contentious History of DNA Fingerprinting* (2008).

economic, and political contexts has been an important theme of socio-legal studies. The next section therefore considers key features of SLS as it pertains to the contemporary study of science and technology.

## SOCIO-LEGAL STUDIES

Socio-legal studies is a heterogeneous field that comprises wide-ranging inquiries into the operation of law in its social, economic, and political contexts. Some contributors to the field apply a 'law-in-action' perspective drawing on empirical data that show how actual legal practices differ from statutory provisions and precedent, 'the law in the books'.[23] SLS's interest in relationships between law and politics overlaps with critical legal studies which seeks to deconstruct formal liberal legal doctrine and reveal its masking of socio-economic, racial, and gender inequalities.[24] SLS encompasses sub-fields such as law and economics, anthropology, history, psychology, sociology, and, more recently, law and geography.

While being concerned with the particular contribution of law – widely defined – to intersections between science, technology, and society, it is worth bearing in mind that SLS does not necessarily argue that law is central in these intersections.[25] In fact SLS has raised the spectre of the rule of technology replacing the rule of law and thereby potentially undermining the role of moral communities in legal regulation.[26] Law has also been perceived as ineffective because it often lags behind technology innovation,[27] paralleling STS's analysis of regulatory lag mentioned above. The possibility of tackling this by drafting 'technology-neutral' legislation has been questioned.[28] Moreover, whether law's regulatory capacity can be enhanced

---

23 S. Wheeler, 'Socio-Legal Research' in *The New Oxford Companion to Law*, eds. P. Cane and J. Conaghan (2008) 1098. This 'gap' analysis has been questioned in socio-legal research that examines creative legal advocacy as a means to close gaps between formal legal rules and actual practices of those who are subject to legal regulation. See, for example, D. McBarnet and C. Whelan, 'The Elusive Spirit of the Law: Formalism and the Struggle for Legal Control' (1991) 54 *Modern Law Rev.* 848–73.

24 P. Fitzpatrick and A. Hunt, 'Critical Legal Studies: Introduction' (1987) 14 *J. of Law and Society* 1–3; D. Kennedy, *Sexy Dressing Etc.* (1995); M. Kelman, *A Guide to Critical Legal Studies* (1987).

25 A. Murray and C. Scott, 'Controlling the New Media: Hybrid Responses to New Forms of Power' (2002) 65 *Modern Law Rev.* 491–516.

26 R. Brownsword and K. Yeung, 'Regulating Technologies: Tools, Targets and Thematics' in *Regulating Technologies,* eds. R. Brownsword and K. Yeung (2008) 3–22, at 6; L. Lessig, *Code and Other Laws of Cyberspace* (1999).

27 The problem of 'regulatory disconnection', Brownsword and Yeung, id., p. 26. T. Ofek, 'Regulatory Dilemmas of Advanced Telecommunications Technologies: The Case of VoIP' (2011) D.Phil thesis, Oxford University.

28 Moses, op. cit., n. 14.

by injecting scientific knowledge or technology norms into law[29] has been queried.[30]

Typically, SLS had a distinct approach to 'the social' as mediating intersections between law, science, and technology. 'The social' was seen as a source of 'meaning making', key to understanding how legal provisions for control of the application of scientific knowledge and technological innovations are interpreted, in particular where the law confers discretionary powers upon regulators.[31] This becomes evident in SLS accounts, for instance, of legal concepts such as pollution[32] or the interpretation of legal rules for regulating waste-treatment technologies.[33] SLS has also examined how 'the social' intrudes into scientific and legal reasoning by questioning the positivist distinction between facts and values. Even quantitative risk assessment has been considered to be informed by value judgements.[34] Thus, SLS's approach to values provides a critique of legal provisions that seek to differentiate sharply between scientific and political approaches to risk regulatory decision making.[35]

Moreover, some SLS studies advocate retaining a political dimension to legal decision making about risky technologies. 'Facts' cannot and should not replace 'values', in particular, the value of participatory decision making.[36] There are limits to scientific knowledge and risk assessment in determining technology choice. Risky technologies also have to be acceptable to citizens. SLS research has pointed out that values, including a lack of attention to issues of social justice, intrude into legal frameworks, such as intellectual property law, that regulate innovative bio-technologies, for example, restricting access of farmers in developing countries to technological innovations.[37] A focus on 'the social' as a source of meaning is also evident in research that examines the incursion of an emotional dimension into the

---

29 Such as a duty imposed through environmental legislation on operators of polluting processes to reduce pollution through the application of 'the best available techniques'.
30 B. Lange, *Implementing EU Pollution Control: Law and Integration* (2008).
31 B. Hutter, *Regulation and Risk: Occupational Health and Safety on the Railways* (2001) ch. 5; K. Hawkins and J. Thomas (eds.), *Making Regulatory Policy* (1989).
32 K. Hawkins, *Environment and Enforcement: Regulation and the Social Definition of Pollution* (1984).
33 B. Lange, 'Compliance Construction in the Context of Environmental Regulation' (1999) 8 *Social and Legal Studies* 549–67.
34 K. Shrader-Frechette, *Burying Uncertainty: Risk and the Case Against Geological Disposal of Nuclear Waste* (1993) 37–8; K. Shrader-Frechette, *Technology and Human Values* (1996).
35 See, for example, Art. 6 (risk assessment) and Art. 7 (risk management) in Regulation 1829/2003/EC on genetically modified food and feed, 18 October 2003, OJ L 268/1.
36 S. Rayner and R. Cantor, 'How Fair Is Safe Enough? The Cultural Approach to Societal Technology Choice' (1987) 7 *Risk Analysis* 3–9, at 3.
37 K. Shrader-Frechette, 'Property Rights and Genetic Engineering: Developing Nations at Risk' (2005) 11 *Science and Engineering Ethics* 137–49.

regulation of risks,[38] advocating, for instance, explicit acknowledgement of emotional content in regulation.[39] Trust, for instance, is analysed not as a source of bias, but as fundamental to facilitating regulation of innovative technologies such as biotechnology.[40] Thus, SLS's contribution is to foreground a distinct, sometimes interpretivist, meaning-making conception of 'the social'. SLS also perceives interests and, thus, strategic action as to be explained with reference to a 'social sphere', with social actors being driven by internal motivations, sometimes shaped by social structures.

Aside from the 'social', however, materiality also matters in SLS. First, some SLS studies invoke a materialist conception of scientific knowledge through their critique of 'technocracy' in risk regulation, construing scientific knowledge as a hegemonic, monolithic force that depoliticizes regulation and marginalizes citizens' perspectives.[41] Second, a recent strand of SLS scholarship about 'law and geography'[42] analyses how law constructs the physical material of space, for instance, through zoning in United States planning law, but also how law is constructed by geographical space, for instance, when legal obligations in relation to water pollution are scaled to a river basin.[43]

Finally, SLS has provided an important contribution to understanding intersections between law, science, technology, and society by highlighting law as a bounded and distinct technique of social ordering,[44] with its own particular processes, techniques, modes of socialization, culture, and belief systems.[45] In this it diverges somewhat from key approaches of STS, as we discuss below.

---

38 B. Lange, 'Foucauldian Inspired Discourse Analysis: A Contribution to Critical Environmental Law Scholarship? in *Law and Ecology: New Environmental Legal Foundations*, ed. A. Philippopoulos-Mihalopoulos (2010) 45, 49.
39 S. Bandes, 'Emotions, Values, and the Construction of Risk' (2008) 156 *University of Pennsylvania Law Rev.* 422.
40 P. Taylor-Gooby, 'Social Divisions of Trust: Scepticism and Democracy in the GM Nation Debate' (2006) 9 *J. of Risk Research* 75–95.
41 T. Hervey, 'Regulation of Genetically Modified Products in a Multi-Level System of Governance: Science or Citizens?' (2001) 10 *Rev. of EC and International Environmental Law* 321–33; M. Lee, *EU Regulation of GMOs* (2008).
42 See, for example, J. Holder and C. Harrison, *Law and Geography* (2003).
43 See the EU Water Framework Directive (2000/60/EC), OJ L 327, 22 December 2000, 1–73.
44 It shares this with some strands of STS research, see, for example, S. Jasanoff, 'Law's Knowledge: Science for Justice in Legal Settings' (2005) 95 *Am. J. of Public Health* S49–58, at S51–2. There is significant variation within SLS. From a systems-theoretical perspective, law forms part of wider communication processes but is an autonomous sub-system (A. Philippopoulos-Mihalopoulos, *Niklas Luhmann: Law, Justice, Society* (2010) 71). But there is a rich strand of SLS that sees law and politics as closely interlinked: see, for example, A. Sarat and S. Scheingold (eds.), *Cause Lawyers and Social Movements* (2006); A. Sarat, *Dissent in Dangerous Times* (2005).
45 See, for instance, J. Paterson, 'Trans-science, trans-law and proceduralisation' (2003) 12 *Social and Legal Studies* 523–43.

# WHERE SLS AND STS MEET: SOCIO-MATERIALITY AND REGULATION

This volume suggests that a deeper understanding of intersections between law, science, technology, and society can be developed by considering synergies between the partly distinct approaches of SLS and STS. More specifically we argue that much of SLS's and STS's insights coalesce around themes that we refer to as 'socio-materiality' and 'regulation'.

## 1. *Socio-materiality*

Both SLS and STS have been influenced by social constructivism and co-constructivism. In STS, such approaches have been applied, for example, in a powerful set of 'controversy studies' that show how conflicts over scientific knowledge or the accounting for technological problems are pursued and resolved through 'social' processes such as interests, reputation, and rhetoric.[46] In SLS, social constructivism has often involved significant reliance on interpretivist meaning making as key to understanding intersections between law, science, and technology.[47] Similarly, the SCOT programme contributed the insight that the design of technologies is subject to 'interpretive flexibility': dependent on the action and reactions of technology users and others, everyday artefacts could have been different.[48] In contrast to SLS, STS perspectives such as ANT remain more detached from the subjective construction of meaning by selected social actors. ANT pays close attention to the material practices of scientific investigation and the operation of technology, by tracing the formation of networks of human and non-human forces. The credibility of scientific knowledge or technology is regarded as a function of the cohesiveness of these networks, and their ability to enrol elements through which information flows.

These partly overlapping, partly distinct methodological influences have shaped how STS and SLS approach their common interest in socio-materiality. We define 'socio-materiality' as material structures embodying social relations and vice versa.[49] Materiality has emerged as a significant

---

46 For example, see S. Sismondo, *An Introduction to Science and Technology Studies* (2004) ch. 10.
47 K. Hawkins, *Environment and Enforcement: Regulation and the Social Definition of Pollution* (1984); B. Lange, 'Compliance Construction in the Context of Environmental Regulation' (1999) 8 *Social and Legal Studies* 549–67.
48 W.E. Bijker, T.P. Hughes, and T.J. Pinch (eds.), *The Social Construction of Technological Systems: New Directions in the Sociology and History of Technology* (1987).
49 See, for example, Winner's example of the 'racist' bridges over the Long Island Parkway in New York, the low height of which kept buses carrying mainly black citizens from going to a public park: L. Winner, 'Do Artifacts Have Politics?' (1980) 109 *Daedalus* 121–36.

contribution from STS. In particular, ANT typically perceives 'the social' as inextricably intertwined with the material,[50] while in SLS materiality, including in technological applications, is more autonomous. In its 'relational materialism'[51] ANT typically conceives of distinctions such as technological-social as functions of a relational, 'flat' ontology. This can be regarded as a version of 'material semiotics'[52] in which heterogeneous networks of elements 'define and shape one another'. The way in which both human and non-human elements are configured into 'assemblages' determines the effects they exert, and the way in which they are perceived in the wider world.[53] Actions and actors of 'law' play parts that are variable and contingent depending on the particular circumstances of such socio-material worlds.

Material agency is highlighted also in the constitution of socio-material worlds when these are perceived through the concept of performativity.[54] Performativity suggests that networks of interacting elements – social-material, and conceptual – produce the worlds we encounter. Callon has attempted to ground the concept of performativity in a pragmatist philosophy based on J.L. Austin's speech act theory,[55] that understands society as constituted by performances of language.[56] Callon argues that science is performative. In this view, scientific activity produces 'indexical' (self-referring) statements that refer to specific events in time and space. If these statements are to be converted into 'universal' statements of scientific fact, they must somehow *create* the 'world' to which they are relevant. Callon uses the term *socio-technical agencement* to describe this performative process of the co-production of theories and the sociomaterial world(s) within which they can exist. This relational process is *agencement*, the term emphasizing how these heterogeneous constructs possess agency. The performative construction of its own meaning is a crucial part of an *agencement*. To make a theory or technology 'work' across domains of time and space, a whole series of actors and objects need to be co-ordinated.

Following Callon, we can see that science and technology 'require' that the social-material world be arranged in a particular, supportive fashion.

---

50 J. Law and A. Mol, 'Notes on materiality and sociality' (1995) 43 *Sociological Rev.* 279.
51 Sismondo, op. cit., n. 46.
52 J. Law, 'Actor Network Theory and Material Semiotics' (2007), at <http://www.heterogeneities.net/publications/Law2007ANTandMaterialSemiotics.pdf>.
53 B. Latour, *We Have Never Been Modern* (1993).
54 M. Callon, 'Introduction: the embeddedness of economic markets in economics' in *The Laws of the Markets*, ed. M. Callon (1998) 1–57; D. Mackenzie, 'An Equation and its Worlds: Bricolage, Exemplars, Disunity and Performativity in Financial Economics' (2003) 33 *Social Studies of Sci.* 831–68.
55 M. Callon, 'What does it mean to say that economics is performative?' (2006) CSI Working Paper, at <http://hal.archives-ouvertes.fr/docs/00/09/15/96/PDF/WP_CSI_005.pdf>.
56 J.L. Austin, *How To Do Things With Words* (1962).

New technologies born from the laboratory are embedded in this world, partly constituted through legal and regulatory regimes, which may or may not be prepared for them and can either endorse or oppose them. Thus here, technology becomes both a subject and a driver of the evolution of law and its varying jurisdictions. The indeterminacies of science, technology, and law cannot be explained or managed by recourse to autonomous 'social' interests alone. The material dimensions of the LSTS nexus also have to be considered. Socio-materiality thus draws our attention to the production of durable relations between law, science, technology, and society, while allowing that they can change at specific historical points. Thinking about socio-materiality therefore requires us to address how intersections between LSTS produce regulatory effects in specific instances, as this volume illustrates.

## 2. Regulation

In approaching 'regulation' both SLS and STS have built on Michel Foucault's work on disciplinary power, discourse, and governmentality. For instance, both SLS and STS have been influenced by Foucault's 'capillary' conceptualization of power as dispersed, immanent in, and constitutive of social action.[57] Key Foucauldian ideas, such as surveillance, discipline, and power/knowledge are of obvious relevance to how science and technological applications produce 'subjects', as in the case of policing through CCTV cameras.[58]

SLS studies have explored how law, science, and technology are discursively constructed, intersect as discourses, and give rise to particular governmentalities.[59] Within STS, an actor-network approach has built on Foucault's understanding of power by developing the idea of hybrid networks. Power is thus enacted through different elements in a network, in which enactors of 'law' will figure to various degrees from case to case. ANT portrays power as an effect generated in hybrid networks of 'actants' that comprise both humans and non-humans as agents. ANT's networked understanding of law also is a valuable development in light of the debate about whether Foucault takes state law seriously enough. Some suggest that he chronicles a historical shift from sovereign state law to 'disciplinary' regulation, but others argue that his analysis of governmentality and

---

57 M. Foucault, *Power: The essential works of Michel Foucault 1954–1984* (2002, 3rd edn.).
58 M. Foucault, *Discipline and Punish: The Birth of the Prison* (1991); D. Neyland and B.J. Goold, 'Where next for surveillance studies? Exploring new directions in surveillance and privacy' in *New Directions in Surveillance and Privacy*, eds. B.J. Goold and D. Neyland (2009).
59 K. Bäckstrand and E. Lövbrand, 'Planting Trees to Mitigate Climate Change: Contested Discourses of Ecological Modernisation, Green Governmentality and Civic Environmentalism' (2006) 6 *Global Environmental Politics* 50–75.

'government of the self' assumes the existence of the sovereign state and its legal force.[60] From an ANT perspective, whether state law is seen in the foreground or background is not fundamental.

One way in which law interacts with scientific or technological materials in networks is through regulatory standards and standardized objects. In Latourian terms, these are 'immutable mobiles'. Material artefacts, such as a periodic table, can provide checks and balances on the exercise of power, and travel to different locations without changing and thereby contribute to the maintenance of order.[61] The diversity of such 'immutable mobiles' is becoming increasingly recognized because regulatory standards may represent 'private' governance or may be mandated by state law.[62] Regulatory standards can also be contestable artefacts, represented in material documents, which themselves can be conceptualized as actor-networks.[63] More recently, analysis of the performativity of the materials of legal procedures and law making provide an important contribution to understanding legal documentation in this way.[64]

While socio-materiality and regulation are key reference points for both SLS and STS, SLS[65] and STS[66] scholars and texts have started to intermingle more widely. A key aim of this volume is to highlight the possibilities for further dialogue. Taken together, the studies featured here demonstrate the considerable potential for a continued process of fruitful interchange. Before introducing each of the articles, we highlight themes common to the collection as a whole.

60 A. Hunt, 'Foucault's expulsion of law: toward a retrieval' in *Foucault and Law*, eds. B. Golder and P. Fitzpatrick (2010).
61 B. Latour, 'Drawing Things Together' in *Representation in Scientific Practice*, eds. M. Lynch and S. Woolgar (1990) 19–68.
62 S. Timmermans and S. Epstein, *'A World of Standards but not a Standard World: Toward a Sociology of Standards and Standardization'* (2010) 36 *Ann. Rev. of Sociology* 69–89.
63 L. Prior, 'Repositioning documents in social research' (2008) 42 *Sociology* 821–36.
64 For example, see Latour, op. cit., n. 16; T. Scheffer, 'File work, legal care, and professional habitus – An ethnographic reflection on different styles of advocacy' (2007) 14 *International J. of the Legal Profession* 57–81; A. Faulkner, *How Law Makes Technoscience: The Shaping of Expectations, Actors and Accountabilities in Regenerative Medicine in Europe* (2010) CSSP Electronic Working Paper No.1, at <http://www.jnu.ac.in/Academics/Schools/SchoolOfSocialSciences/CSSP/CSSP-EWPS-1.pdf>.
65 See, for example, B. Lange, 'Social Dynamics of Regulatory Interactions: An Exploration of Three Sociological Perspectives' in *Law and Sociology*, ed. M. Freedman (2006) 141–64; T. Scheffer and K. Hannken-Illjes, 'Courtooms de-centred. Comparing the English Crown Court and the German Landgericht' (2007) 28 *Zeitschrift für Rechtssoziologie* 229–39.
66 See, for example, E. Cloatre, 'Trips and Pharmaceutical Patents in Djibouti: An ANT Analysis of Socio-Legal Objects' (2008) 17 *Social and Legal Studies* 263–81; E. Stokes, 'Regulating Nanotechnologies: Sizing Up the Options' (2009) 29 *Legal Studies* 281–304, fn. 82.

## THE CONTRIBUTION OF THE ARTICLES TO UNDERSTANDING SOCIO-MATERIALITY AND REGULATION

The articles in this volume address several broad, key questions:
(i) How do law, science, technology, and society interact to bring into being new socio-material realities? Why do some realities become dominant and what part do legal processes play?
(ii) How are standards, measurement regimes, and ethical positions built into law and regulation?
(iii) What are the spatial or geographical dynamics of law?
(iv) How do legal classifications develop, expand or get disrupted, and how do they 'match' the institutions of legal enforcement?
(v) And, most radically, how is law constituted by socio-materiality?

The articles address these questions from the vantage point of specific thematic, geographical, and jurisdictional domains. They deal with biomedicine, social policy, databanks and data protection, the physical environment, criminal investigation, and more, covering the Netherlands, France, the United Kingdom, the European Union, and global jurisdictions. The legal and socio-material aspects of fundamental dynamics of everyday life, and indeed 'life itself'[67] are raised in the articles, through themes of social and personal identity, social inclusion and exclusion, power and authority, human rights, citizenship, privacy and family, as well as public order.

The first of our articles sets the scene by providing a fascinating glimpse of a rarely-seen, but nonetheless instrumental site of synthesis between law, science, and the wider socio-material world. Javier Lezaun presents an ethnography of the Institute for Reference Materials and Measurements (IRMM), which is tasked with producing a wide and esoteric array of substances, required to ensure European regulations have standardized foundations in the real world. Lezaun's study shows how ensuring unbroken chains of correspondence between these reference materials and their comparators is a complex affair, involving a high degree of ingenuity to manipulate substances which can function as reliable, inert material referents for standards. Giving examples such as materials testifying to marine pollution, Lezaun shows how such complexity is largely hidden from the gaze of European law-making institutions, but is absolutely vital for making sense of the practices of regulatory judgement.

Latour's recent study of decision making in the Conseil d'Etat[68] represents a new departure with its focus on the transactions of people, reasoning, materials, and texts that enact law. With notable exceptions,[69]

---

67 N. Rose, *The Politics of Life Itself: Biomedicine, Power, and Subjectivity in the Twenty-First Century* (2006).
68 Latour, op. cit., n. 16.
69 An early example is A. Riles, *The Network Inside Out* (2000).

ANT, with which he has often (sometimes reluctantly) been associated, has remained relatively unexplored for understanding the evolution of new regulatory regimes. Rooke et al. seek to redress this in their study of the emergence of medicinal-nicotine (MN) products. They utilize an ANT approach to describe the challenges which nicotine gum presented to regulatory frameworks within European jurisdictions. Their study shows how nicotine gum had a highly ambiguous status in the context of regulations at the advent of MN products. Nicotine gum challenged previously held understandings of 'medicines', and Rooke et al. chart how a complex series of transactions were required to enrol it into regulatory frameworks. Their account also highlights the fluid nature of regulation, and how the process of enrolment itself created new regulatory networks which, in turn, generated new socialized understandings of regulatory categories.

Actor-network approaches are sometimes criticized for being 'apolitical'. Ilke Turkmendag examines the way in which the science of genetics has entered into the technologies, materials, and social structuring of human procreation and the legal-political actors (and activists) who have involved themselves in 'political' debate. Technologically-aided reproduction has become relatively commonplace for creating children and building families. The material aspects of this practice lie in matters of tactile objects such as gametes (eggs and sperm). Turkmendag analyses the radical shift in the United Kingdom from a position of anonymous donation of sperm to the 'child's right to know' its sperm donor, introduced in 2004. She tracks the contrasting discourses of stakeholders in court cases, government debate, and in the public media, showing how a discourse of a 'right to know' genetic inheritance prevailed. It is interesting to note that, although issues of disease inheritance were raised, they did not figure significantly. Turkmendag shows how appeals to analogy with child adoption law was crucial in the change to legislation. She concludes by questioning this shift: shouldn't society treat the intending 'would-be parents' – rather than the gamete donors – as those legitimately providing the social identity of the child?

Turning from national to global regulation, and from the human biological to the natural, physical environment, Emilie Cloatre and Nick Wright discuss carbon emissions in the context of climate change. In particular, they examine the 'Clean Development Mechanism' (CDM) of the Kyoto Protocol. The authors draw on Michel Callon's notion of 'engineer-sociologists' shaping 'actor-worlds' to show how the complex actors involved in 'movements of failure, transformation and resistance' to the CDM necessarily co-construct accompanying visions of society. Some of these visions appear more compatible with the design of carbon-emission projects than others. Population-based projects seem less amenable to the technicalities and bureaucratic accountabilities of the CDM than more easily quantifiable single-site technologies. The authors provide an analysis of how these materialities built into the CDM produce or replicate social divisions.

They show how a commensurating legislative script, that seeks to 'make things the same',[70] in practice produces likely winners and likely losers.

The emerging materials of nanotechnology show that 'size matters' in law. Stokes's article traces the development of EU legislation where there has been a clear case of regulatory lag. Stokes examines the way in which EU regulators have tried to stretch the 'inherited regulatory environment' of consumer products to cover nano-products. In spite of some principled opposition to nanotechnology and calls for specific legislation, the approach adopted focuses on information provision via labelling of products. Existing chemicals legislation defines 'substance' in a way that can be taken to include nano-materials, and Stokes shows how the complexity of existing consumer rights legislation has been significant. The EU promotes a vision of the need for 'safe' and 'confident' consumers, tied to the economic aims of the EU polity, showing the EU's stake in the 'everyday reality in the lives of its citizens'. Stokes critiques the analogy between conventional bulk-scale products and nano-products, showing that a 'nano' label may be obfuscating, and questioning whether citizens' concerns about a world strewn with nano-products can be met without public involvement and better data.

The importance of disputable legal analogy is also shown in a contribution on 'biobanks'. The collection of human materials in the form of tissues, cells, and information in biobanks is becoming one of the cornerstones of societies' resources for disease research, biomedical innovation projects, and building of national imaginaries. Emmanuelle Rial-Sebbag and Anne Cambon-Thomsen discuss biobanks as another case challenging the extension of a regulatory heritage, critically analysing how EU law fails adequately to cover the sociomaterial realities of biobanking. The authors locate biobanking in the predecessor legislation on biotechnology; analyse the significance of the rights of 'source-persons'; and outline the development of European bioethical principles. They argue that 'legal analogies between the human being, the samples from the body, and the rights of human beings' encounter problems with the 'semi-permanent' link between person and sample. As the EU starts to develop transnational biobanking, they discuss how citizen engagement projects may help provide the basis for stronger and more inclusive governance systems.

The sheer complexity of the classificatory processes of legislative regulation in the human tissues and 'regenerative medicine' field is illustrated in Aurelie Mahalatchimy et al.'s contribution, again with an EU focus. Their analysis first seeks to account for complexity by contrasting implementation in France and the United Kingdom. Is this complexity simply the translation of the inherent biological complexity of human material uncovered by science? Secondly, the article examines the 'match' between legal provisions

---

70 D. MacKenzie, 'Making things the same: Gases, emission rights and the politics of carbon markets' (2009) 34 *Accounting, Organizations and Society* 440.

and regulatory institutions. The EU has invented a category of 'advanced therapy medicinal products', not used elsewhere in the world and has located this regulation within the pre-existing EU pharmaceutical regime. Comparing the process of 'non-industrial' production of therapeutic products in hospitals under this regime lays bare major differences in the ethical basis of contrasting national regimes in France and the United Kingdom.

While many of the articles concern the regulatory dimensions of emerging technologies in various fields, they also show that law itself increasingly depends upon contributions from scientific advances. Nowhere is this more apparent than in criminal justice. DNA technology has had a major impact on criminal justice across many jurisdictions, and is the focus of Victor Toom's article. Toom shows how science and technology increasingly shape understandings of the 'suspect' individual. His article describes how science and law interact to create new relationships of control between the individual and the law, and new investigative strategies, via the construction of 'forensic genetic bodies'.

Finally, Alain Pottage poses a radical, subversive question: the materiality of what? Pottage provides a timely theoretical complement to the empirical studies we present. In particular, Pottage takes issue with Latour's treatment of law, contrasting it with the same author's celebrated studies of science and technology. The key problem, he argues, is that Latour's characterization of law as a 'regime of enunciation' precludes any meaningful synthesis with the material framing of sociality. In considering strategies for bridging this gap, Pottage also finds the influential work of Luhmann to be wanting, but identifies possibilities for reconciling a materialist approach with the concept of the *dispositif*, as advanced by Michel Foucault.

Overall, the articles in this volume share a number of common themes, and they develop innovative directions in order to better understand intersections between law, science, technology, and society, The first commonality lies in the concept of 'socio-materiality' itself. Whether we study the material (scientific or technological) basis of law or its translation into practice, we find the malleabilities of the social and of the material to be interdependent. In these contributions we see the extremely dynamic nature of the relationship between law, science, technology, and society, in so far as we wish to separate these categories. Two ostensibly counterposed phenomena are apparent. Technology can be seen to be both an object of law, and as a means (sometimes unintended) of engendering new laws and legislative understandings. In some cases technology may advance new relations of power, such as in Toom's example of forensic DNA carving out a new, problematic relationship between the individual and the state. Yet what also emerges from the collection is a sense in which technological and legal orderings frequently *co-produce* each other.[71] In many cases, the two

71 Jasanoff, op. cit., n. 18.

dynamics occur interdependently, a phenomenon which Rooke et al. and Mahalatchimy et al. highlight with particular clarity.

The articles do, however, vary in their approaches to law and in the way in which they point to new directions. Some focus more upon law as an arena of discourse *about* forms of materiality and material worlds (Rial-Sebbag/ Cambon-Thomsen; Stokes). Others focus more on the precariousness of law's material foundations (Lezaun), or on the social stratifications of its socio-material translation into practice (Toom; Cloatre/Wright). Others strike various balances between the legal and the socio-political (Turkmendag), the legal and the social-institutional (Mahalatchimy), and the legal in socio-material networks (Rooke) – if we accept the constitutive, power-invoking nature of such categorizations (Pottage).

At the heart of the socio-material domains illuminated in this volume, furthermore, we see a tension between stabilization through, on the one hand classification, categorization, measurement, standardization, commensuration,[72] and analogy, and, on the other, disruption, negotiation, elasticity, limitation, escape, resistance, and reformulation. Classification is a fundamental meaning-making social activity,[73] key to the workings of law, and here we see the materiality of science and technology appearing as both classifying actor and as classified object. Several of the analyses show the key dynamic of actors wrestling with the path-dependence of existing regulatory regimes, 'inherited regulatory environments' (to use Stokes's term), sometimes resulting in new or extended regimes, sometimes not. The extension of the material world through biomedical innovation or carbon emission testing challenges the limits of law, regulation, and governance. Not only social actors – 'stakeholders' – exert power in reshaping the legal landscape, but law itself shows a material obduracy in its textuality and through the socio-legal categories it mobilizes to structure society. Analogy is a well-known tactic in law making and adjudication, but here we see how this practice of commensuration, of 'making things the same',[74] can be contested and constructed through discourses of the powerful or resourceful, and how some scientific and technological artefacts obstinately resist or otherwise shape commensuration projects.

---

72 W.N. Espeland and M.L. Stevens, 'Commensuration as a social process' (1998) 24 *Ann. Rev. of Sociology* 313–43; W. Orlikowski, 'The duality of technology: rethinking the concept of technology in organizations' (1992) 3 *Organization Sci.* 398–427, at 405, 421.
73 G. Bowker and S.L. Star, *Sorting Things Out: Classification and its Consequences* (2000).
74 Mackenzie, op. cit., n. 70.

Given the diversity of scientific and technological development, and the varied ways with which they interact with law and society as a whole, one collection alone can capture only a limited range of empirical topics. We believe, however, that the articles here provide a rich array of insights to provoke further investigations and understandings of the intersections of law, science, technology, and society. Continued theorizing about LSTS will further develop the conceptual tools displayed and discussed in this volume. Such tools will have to be able to address relationships between law and science and technology of ever-increasing complexity.

Substantive areas not covered by this special issue certainly demand further attention. In particular, technologies which promote interconnectivity across jurisdictions present opportunities for further research. Information and communications technologies (ICT) represent a key area, as both an object and tool of law making, raising questions about the perceived use and abuse of ICT. Controversies such as file-sharing and the protection of copyright in social media will remain a priority.[75] The way in which ICT is increasingly shaping processes of law making and law enforcement has only begun to be considered in depth.[76] The rise of forensic computing, for example, is an important development. In terms of the theoretical focus of this volume, ICT raises particularly challenging issues of identifying 'materiality', and offers opportunities for further elaboration of this concept as the world becomes increasingly dependent upon, and constituted by, electronic communication.

Other kinds of critical infrastructures present further possibilities. Transport infrastructures, for example, are loci where technological, social, and environmental factors intermingle in complex ways, with significant risks at stake. Transport regulation has recently come under the spotlight due to events such as the 2010 Icelandic volcanic ash crisis, which added fuel to debates over the regulation of European airspace management. The tensions between proposals for the integration of European airspace and existing legally enshrined norms of airspace governance, is a pressing issue.[77] This is another example of how the construction of particular LSTS intersections has an immediate impact on the daily life of millions of people. Many other examples could be added.

75 J. Hoffman and S. Botzem, 'Transnational governance spirals: the transformation of rule-making authority in Internet regulation and corporate financial reporting' (2010) 4 *Critical Policy Studies* 18–37.
76 For example, A. Vranaki, 'Law as a Manifestation of Power in Online Social Networking Sites: An Analysis of the Protection of Privacy and Copyright Interests' (2012, forthcoming) D.Phil thesis, Oxford University.
77 A. Allemano, 'The European Regulatory Response to the Volcanic Ash Crisis: Between Fragmentation and Integration' (2010) 1 *European J. of Risk Regulation* 101–6.

At the same time, however, other avenues for research reflect more longstanding SLS and STS concerns. For example, while many of our authors provide a welcome focus on the supranational dimensions of regulation, more needs to be done to understand the way in which techno-scientific change impacts upon legal systems outside the Western developed world.[78] The relationship between technology and development has mainly been discussed from an economic perspective, yet there remains considerable scope to explore how technological development challenges social norms across jurisdictions in emerging and developing economies. The way in which norms and understandings of human rights interact with new technology is just one topic inviting further consideration.[79] What consequences might the spread of surveillance technologies and forensic DNA databases across non-Western jurisdictions hold for the rights of the individual subject in relation to the state?

One must exercise caution when considering how the law-science-technology-society relationship will develop, and how it will shape the 'socio-material worlds' that we create and inhabit. The articles here show that intersections between science, technology, law, and society will continue to unfold in complex ways. They also show how emerging technologies challenge our understandings of crucial distinctions – between the local and supranational, between the 'social' and the 'natural', and the material and the ideational. The interstices of the relationships between law, science, technology, and society are becoming simultaneously more complex and more delicate. We hope that this collection provides insight into the opportunities for understanding these developments, for building on them further, and for enriching the academic cross-fertilizations they represent.

---

78 S. Jasanoff, 'Introduction' in *Genetic Suspects: Global Governance of Forensic DNA Profiling and Databasing*, eds. R. Hindmarsh and B. Prainsack (2010).

79 See O. Bekou and T. Murphy (eds.), *Human Rights and New Technologies*, Special Issue (2010) 10 *Human Rights Law Rev.*

# The Pragmatic Sanction of Materials: Notes for an Ethnography of Legal Substances

JAVIER LEZAUN*

*How is the law bound to the material world? This article examines the production of reference materials, artefacts that incarnate legally relevant measurements and serve as a transitional object for the law in its approximation to the stuff of the world. The argument, an opening for an ethnographic investigation into the life of legal materials, is based on a study of the work conducted at the Institute for Reference Materials and Measurements (IRMM), an agency that, for over fifty years, has fabricated official versions of the objects and substances mentioned in European law.*

## INTRODUCTION

How does law come to matter? How does it manage to refer to anything material at all, and thus introduce a new sort of force – a legal force – into the world? By what strange alchemy is a legal obligation transformed into a material constraint? This essay explores a peculiar intermediary between the legal and the material, a point of passage for the law as it bridges the distance that separates it from the world of mundane objects. It deals with the production of *reference materials*, official substances that incarnate legally relevant measurements, transitional objects that offer the law a point of contact with the stuff of the world. The discussion is based on a study of the

---

\* Institute for Science, Innovation and Society, School of Anthropology and Museum Ethnography, 64 Banbury Road, Oxford OX2 6PN, England
javier.lezaun@anthro.ox.ac.uk

Initial fieldwork for this project was sponsored by the Centre for Analysis of Risk and Regulation (CARR) at the London School of Economics. Subsequent research received funding from the European Research Council under the European Community's Seventh Framework Programme (FP7/2007-2013)/ERC grant agreement no. 263447 – BioProperty. I would like to thank the members of IRMM staff who collaborated with my research, and Michael Guggenheim and Ann Kelly for comments on earlier drafts of the article.

work carried out at the Institute for Reference Materials and Measurements (IRMM), a research establishment that, for more than fifty years, has produced the physical referents of European Union law.

A reference material is a physical object with agreed upon qualities – as the International Organization for Standardization puts it, 'a material, sufficiently homogeneous and stable with respect to one or more specific properties, which has been established to be fit for its intended use in a measurement process' – and IRMM is the agency tasked by European authorities with the fabrication of dependably constant versions of the entities mentioned in EU legislation. Located on the outskirts of the Flemish town of Geel, the Institute traces its origin back to the inception of European integration: its forebear, the Central Bureau for Nuclear Measurements (CBNM), was established under the auspices of the 1957 Euratom Treaty. Since then, IRMM, part of the European Commission's Joint Research Centre, has steadily expanded its mission from its original nuclear mandate, and today produces reference materials in virtually all areas of European regulatory law.

Perusing the catalogue of materials available from IRMM is like reading the inventory of a curiosity cabinet of European integration. Among the more than 500 substances listed there, one finds bovine eyes positive for clenbuterol (Certified Reference Material BCR-674: vials containing eye liquid from cows fed on the chemical), and toasted bread with acrylamide (ERM-BD273: bread powder in amber glass bottles); a reference for 'urban dust' (BCM-605: used to detect the use of chemical compounds in gasoline), or the recently produced first-ever certified nanoparticle (ERM-FD100: glass ampoules with silica particles suspended in water certified for size).[1]

These are all entities mentioned, somewhat casually, in European laws and regulations, but which the law would be unable to find in the world if it were not for the availability of an official, stable version of the material in question. Take for instance the recent legislation on the presence of contaminants in foodstuffs.[2] Among many other things, it forbids the commercialization of milk containing more than 0.020 milligrams of lead per litre. To enforce or comply with this rule, samples of milk are routinely collected and tested for the presence of lead, but given the required degree of confidence in measurement, the chances are that different analytical methods applied to the same sample – or the same method carried out at different locations or by different people – will yield results sufficiently dissimilar to cast doubt on whether the sample in question is positive or negative for

---

1 The different acronyms refer to the programmes under which the different reference materials were produced. BCR stands for Bureau Communautaire de Référence, the umbrella under which most materials were produced in the 1980s and 1990s. ERM (European Reference Material) is a brand introduced in the last decade.
2 Commission Regulation (EC) No. 1881/2006.

excessive lead content. What to do then? The law is unable to discriminate between these contradictory measurements; it cannot adjudicate analytical controversies. Only by means of a certified reference material, an official version of 'milk containing X amount of lead', can instruments be verified and measurements disciplined, thus allowing the legal rule to pass through the inevitable analytical disagreements and acquire force vis-à-vis particular objects and measurements.

Technicians at IRMM sometimes describe their materials as 'truth in a bottle'.[3] In the case of 'milk with lead', the truth in a bottle is Certified Reference Material BCR-063R, a small vial with about 50 grams of milk powder that, IRMM vouches, will for a year contain a precise quantity of lead (18.5 ng/g, with an uncertainty of 2.7 ng/g). Several aspects of this promise stand out: how can a small amount of milk powder (derived, in the case of BCR-063R, from 1,200 kg of skimmed raw milk collected in the Netherlands in December 1990) stand for *any* kind of milk, anywhere in the world? Why the peculiar temporality of the qualities ascribed to the material, the precise duration of IRMM's guarantee? The point that needs to be emphasized from the start, however, is that the circulation of this vial marks the limit of the law's reach. Without a material reference, the imposing Commission Regulation (EC) No 1881/2006, establishing a maximum level of lead in European milk, would remain a wishful declaration with no physical connection to lead, milk, cows or grass. It is BCR-063R, and its use in hundreds of laboratories and thousands of measurements, that give the law its ability to discriminate the material world.

This article approaches the material mediation of legality – and, we will see, the juridical dimensions in the fabrication of reference materials – from four vantage points. First, it explores the forms of paperwork that underpin the fabrication of a certified reference material. Rather than being a physical counterpoint to the textual categories and properties of the law, in a remarkable example of transmutation, the material reference itself emerges out of a complex process of writing. Second, reference technicians often describe their particular expertise as knowledge in 'how to handle stuff'. What makes someone an expert in 'handling stuff', and how is this ability connected to the striking versatility of the IRMM laboratory – a place where

---

3 Throughout the article I will use the term 'reference technician' to encompass what is a very heterogeneous group of scientists and researchers. About a hundred people, with PhDs in a diverse range of disciplines, work at IRMM's Reference Materials Unit. By describing them as 'technicians' I do not mean to diminish their scientific achievement. The term hints, rather, at their relative invisibility in the machinery of European law making. S. Shapin, in 'The Invisible Technician' (1989) 77 *Am. Scientist* 554–63, has described how the invisibility of technical work accompanied the rise of a scientific culture in seventeenth-century England. A similar argument could be made about the invisibility of these technicians of the law (compare B. Latour, *The Making of Law: An Ethnography of the Conseil d'Etat* (2010)).

objective versions of the most diverse natural kinds – from plutonium to herring, peanuts to coastal sediment – are manufactured?

Third, the production of reference materials can be further illuminated by examining how reference technicians pursue the goal of producing ever more 'realistic' versions of the material in question: a physical standard in the form of 'fresh' fish, say, rather than fish powder or fish oil. The artistic dimension of reference materials comes across most clearly in this effort to produce a technical standard possessing a certain naturalistic verisimilitude. Fourth, and finally, the article describes a fundamental tension in the community's understanding of the relationship between the authority of an official substance and its material nature. By pursuing these four threads I hope to suggest some of the work that is involved in making the law mundane, in allowing the law to speak of the world, to be of the world. Paper trails, handling skills, the artistic reconstitution of nature, and the ideal of non-artefactual objectivity are all facets of the labours required to transform law into a force capable of ordering our material world.

## PAPERWORK

'We are the only part of the European Commission that produces something other than paper.'[4] This announcement from the Head of the Reference Materials Unit, a sort of welcoming gambit for the visiting sociologist, is, at first sight, disconcerting, for the offices and laboratories at IRMM are quite obviously full of paper – most of it, it seems, self-produced.

It is true, though, that paper, while central to the day-to-day activities of reference technicians, is not their most notorious product. Pride of place belongs to the bottles, ampoules, and vials, some several decades old, on display in the showcases that line the hallways of the Unit (Figure 1). Each of them contains the result of several years of work: a physical substance or matrix with a certified value for a particular analyte, attesting to a category of European regulation. There is, however, an intricate relationship between writing and manufacture, between texts and artefacts. Writing and documentation cover so seamlessly all the stages in the fabrication and circulation of certified reference materials, that the physical substances manufactured in Geel are in an important sense the distillation of a paper process.

The first and most obvious indicator is the fact that no reference material produced at IRMM leaves its premises without the company of a set of texts and documents – or, rather, if it were to leave IRMM without its paper escort it would no longer be a *reference* material, but automatically be transformed

---

4 Quotations attributed to members of the Reference Materials Unit, and not accompanied by a bibliographical citation, are drawn from fieldwork interviews conducted at IRMM in 2010 and 2011.

**Figure 1. Showcases with samples of reference materials in the new IRMM production facility. The materials are often accompanied by images or objects representing the worldly things and substances from which they originate (a cob of genetically modified maize, mussel shells, chemical drawings of the analyte, and so on).**

Image courtesy of IRMM. © European Union, 2011

into another worldly substance. The key piece of paper that travels with the reference material is its Certificate. Signed by the Head of the Reference Materials Unit and printed on European Commission and IRMM letterhead, it lists the substance in question ('human blood', 'natural Moroccan phosphate rock', 'sewage sludge from industrial origin', 'synthetic wine'), and the certified value, with their respective uncertainty estimates, for a series of analytes or properties.

The Certificate for Human Blood (BCR-635), for instance, describes the sample ('[L]yophilized human whole blood in brown glass vials each containing approximately 0.6 g dry matter with a residual moisture content of less than 2% and equivalent to 3.0 mL of fresh whole blood. Sodium-EDTA was used as anticoagulant'), lists the values for the mass concentration of cadmium (6.6 $\mu$g/L – uncertainty 0.6 $\mu$g/L) and lead (210 $\mu$g/L – uncertainty 24 $\mu$g/L), and specifies the period of the Certificate's validity (one year after purchase). As we noted, every reference material comes with a precise 'shelf life' – IRMM will only guarantee the values stated in the Certificate for a specific period of time. After a year, and despite the protection of the glass vial, the interaction of the material with its environment could have led to sufficient instability to compromise the certified properties of the reference.

On the back of the Certificate, the user will find a list of the analytical methods used for measurement, and the names of the institutions that participated in the certification of the material. In addition to monitoring the extraction and processing of the source material, IRMM coordinates the inter-laboratory comparison through which the agreed certified value is produced. The Certificate also includes safety information, storage recommendations ('Upon arrival, the material should be stored at $-20^\circ$C or lower for not more than 12 months until use'), and, finally, instructions for use (for example, 'Allow the vial to reach ambient temperature before opening', or 'Tap the bottom of the vial to loosen any blood material adhering to the stopper').

The Certificate is accompanied by an even more densely packaged document, the Certification Report, which describes in detail how the reference material was produced, from the moment the raw substance was collected ('Thirty-five donations of fresh blood were obtained from normal (healthy) Danish blood donors in October 1996'), to the statistical analysis of the measurements obtained in the inter-laboratory comparison. The Report summarizes the work that went into producing a suitably homogeneous and stable material, and circumscribes the analytical uses to which it can be put.

In combination, Certificate and Report play a critical role in conferring the physical artefact a referential value. They are performative, in the sense that they establish the value of the material in question as an *official* reference. Yet it is important to note that these documents are the outcome of a much more extended process of writing and documentation, a process that precedes the initial sourcing of the worldly substance and runs parallel to its transformation into a referential material. In the course of manufacturing a reference material, the offices and laboratories at IRMM produce large amounts of paper, compounded by the amounts produced by the other laboratories that participate in the certification campaign. The Certificate and the Report thus draw together a multitude of private texts and inscriptions, produced in dozens of locations (fifteen institutions participated in the fabrication of BCR-635), over several years of work. It is impossible to discuss here the multiple paper trails generated in this effort, but it is useful to describe briefly the types of documents that control the working routines of IRMM technicians.

Work at IRMM is governed by a hierarchy of texts. At the top of that hierarchy is the *Quality Manual*, a document that lists the general principles that ought to govern the production of reference materials. The next level of writing is the description of the *Activities* carried out at IRMM, general categories of practice – calibration, testing – and the conditions of their successful accomplishment. Activities are in turn composed of *Procedures*, a total of fourteen, which prescribe a goal to be achieved and describe a standard approach that usually works to achieve such goal. Finally, *Working Instructions* describe, in a step-by-step fashion, singular tasks, such as 'Cleaning of glass penicillin vials for storage of large-scale dried spikes', or

'Separation and purification of uranium for measurement of isotopic ratios by TIMS'. The fabrication of a single reference material can in some cases be governed by more than a hundred of such Instructions, and, significantly, these documents are always written by the members of staff working on that particular reference (whereas texts higher up in the hierarchy of documentation are drafted and updated centrally). There are, moreover, many general practices of material processing, such as milling or sieving, which have no written description, for they are considered too specific to the source material in question to merit a write-up.

All this writing serves the double purpose of instructing technicians in the appropriate routines and principles of reference material production, and of documenting the work carried out at IRMM so as to demonstrate to outsiders the 'administrative objectivity' of the process through which referential products were fabricated, the alignment of institutional practices with a set of ideal-typical bureaucratic procedures.[5] Constant writing underpins IRMM's formal accreditation as a certifier of reference materials. If certification is 'an attestation of compliance', whereby a practical activity is demonstrably shown to have been carried out according to a pre-established set of rules, resulting in a product that can be said to be obedient to a particular understanding of quality, accreditation is an 'attestation of competence', the corroboration of a capacity to perform at a certain standard. The accreditation of producers of reference materials is governed by the rules of the International Standards Organization (ISO Guide 34), and administered by the International Laboratory Accreditation Cooperation (ILAC). IRMM was one of the first establishments accredited for the production of a wide range of reference materials and analytes, and that accreditation rests largely on a series of documentation practices. The signature of the Head of Unit in every Certificate is thus backed by many other signatures – those of the auditors attesting to the reliability of IRMM as an organization able to manage its working processes according to the relevant international rules.

This continuous effort to describe in writing the activities of processing and fabrication reflects the fact that a reference material is not merely a physical artefact, but also a legal device; documentation is a means of strengthening that material in the event of a legal analysis of its authority. As Mallard notes in his study of legal metrology, practices are 'made explicit and accountable partly because of the juridical issues at stake'.[6] Writing is, moreover, a way of compensating for the irreversibility of the transformations effected at IRMM, a means of memorializing the changes a substance went through on its journey to become an official standard. The Report that accompanies every reference material tells a tale in which all the actors that

---

5 M. Lynch, S.A. Cole, K. Jordan, and R. McNally, *Truth Machine: The Contentious History of DNA Fingerprinting* (2008) 136 ff.
6 A. Mallard, 'Compare, Standardize and Settle Agreement: On Some Usual Metrological Problems' (1998) 28 *Social Studies of Sci.* 571–601, at 578.

participated in its production can have their part, particularly those that disappeared or were destroyed in the process. The bovine eye liquid with a particular concentration of clenbuterol (BCR-674) is no longer traceable to any cows ('The eyes were taken with gloves instantly after death, stored by pairs on ice and brought to the laboratory'); the material used to fabricate BCR-605 ('urban dust') was collected at a particular time, in a particular place (the Queensway road tunnel in Birmingham city centre, in February of 1994) and no samples of the original stuff have been kept.[7] The Report that describes the production of each material is thus a sort of *biography*: as with any biography, we learn in it snippets about the material's birth (the cattle were male and female German Simmental, weighing between 450 and 600 kg each; the road dust, about 15 kg, was collected by sweeping a lay-by area in the central section of the Queensway tunnel), but the bulk of the narrative is an account of how the material in question became an authoritative artefact.

Uninterrupted writing is thus the means by which the work carried out at IRMM is planned, documented for the benefit of outsiders, and narrated to those who were not present when the original substances were collected or transformed. If the standard laboratory executes routinely the mesmerizing task of 'transform[ing] pieces of matter into written documents', as Latour and Woolgar famously showed in their ethnography of the Salk Institute,[8] the Reference Materials Unit performs the even more astonishing feat of turning written documents into a piece of matter, of distilling a material reference out of a mass of text.

### 'CERTAIN WAYS OF HANDLING STUFF'

Writing pervades the activities of reference technicians, yet when pressed to describe the form of expertise that characterizes this heterogeneous community, the most common answer is that they are all experts in *handling*. They do not think of themselves as accomplished writers, which they obviously are, but as craftsmen in the arts of manual processing. In the words of the Unit Head: 'People [here] are used to certain ways of handling stuff'.

'Certain ways of handling stuff' might not seem a very specific way of describing a professional or expert domain, but it is precisely this lack of

---

7 Source material, in its worldly form, is never kept – it would be too unstable to offer any kind of referential support. What is common at IRMM is to produce 'reference samples' of a reference material. These are samples of the finished artefact that are stored under more protective conditions than the normal stock of the reference material (for instance, at lower temperatures). They can then be used to analyse the relative degradation of the normal stock of the standard.
8 B. Latour and S. Woolgar, *Laboratory Life: The Construction of Scientific Facts* (1986) 51.

specificity that characterizes reference technicians and their work. Their ability to work across a wide range of materials, from isotopes to peanut butter, urban dust to mouse brains, combined with a heightened attentiveness to what is distinct and idiosyncratic about each substance, is arguably what sets this community apart from other specialized scientific groups.

What does this expertise in *handling stuff* consist of? Reference technicians are experts in the methods and techniques that may be employed to render matter *inert* – at least as inert as possible, and for as long as possible, to allow steady observation across locales. This craft rests on at least three pillars. First, on the mastery of a set of practices, such as drying, grinding, and mixing, used to change the physical state of matter and transform worldly stuff into substances amenable to laboratory observation. These practices might appear generic, but in fact they are so specific to particular materials – not simply to particular natural *kinds*, but to the peculiar origin and history of each of the source materials that enters the laboratory – that, as we saw, they are not even written up.

The second dimension of the reference technicians' craft concerns the provision of protective surroundings. The stillness of matter requires a careful processing of the material, but also the configuration of environments that limit the chances that the material will enter into contact with other materials. These environments include the laboratory itself – its ventilation system, the flow of materials across the different spaces, its equipment (clean rooms, glove boxes, freeze-driers, bottling implements, and so on) or the design and maintenance of dedicated storage facilities – but also the provision of specific atmospheres for individual materials (the packaging format and inert gas most suitable for the relevant kind of matter), and a constant concern with the safety of references in transit: 'one has to avoid weekends for dispatch as this may lead to longer dead-times, try to find courier services willing to replenish dry-ice use for cooling, etc'.[9]

The third pillar of the reference technicians' craft is a deep familiarity with some common implements of material handling. The transformation of lively (and lively inhomogenous) matter into powders, for instance, tends to involve a number of common tools, from the beater-mill used to pre-crush the source material, to the cone and turbula mixers used to homogenize the sample. Most of these implements are purchased from commercial manufacturers; others, like a whole-Teflon jaw used to coarse, are built in-house.

The glove box (Figure 2) – critical in protecting both workers *and* materials from the contamination that inevitably ensues from contact between them – offers a telling example of how a hard-earned ability to operate seemingly simple machinery connects the most diverse kinds of

---

9 T.P.J. Linsinger et al., 'Preservation of sensitive CRMs and monitoring their stability at IRMM' (2004) 378 *Analytical and Bioanalytical Chemistry* 1168–74, at 1170.

**Figure 2. Glove box used for vacuum deposition at IRMM. The glove box has been a constant element in the Geel workshops, from the time of the original Central Bureau for Nuclear Measurements.**

Image courtesy of IRMM. © European Union, 2011

materials and permits the IRMM community to build long-lasting and versatile sets of skills. An oft-told story of IRMM's adaptability goes back to the 1990s, when news of the transmissibility to humans of 'mad cow disease' rocked Europe. IRMM was tasked with the assessment of a test for the detection of bovine spongiform encephalopathy and, in a matter of days, technicians emptied a laboratory dedicated to the analysis of radioactive chemicals, replaced the nuclear materials in the glove boxes with cow brains and spinal cords, and began to produce thousands of samples of bovine tissue for the identification of infected animals.

A relatively stable set of devices explains the peculiar versatility of reference technicians, their ability to work one day on rice flour, and the next on sewage sludge, or synthetic wine. In this way, every reference material is intimately connected with the rest of the creatures fabricated at IRMM: the prion and the peanut butter are not that far apart; the silica nanoparticle shares part of its biography with the bovine liver. They are joined by their passage through the machines of this strange laboratory, where material versions of the law are synthesized.

The manipulations that give rise to an official substance also set it apart from the original state in which it was found in the world. In metrological parlance, all the steps necessary to produce a stable reference material reduce the material's *commutability*, the 'closeness of agreement' between the behaviour and qualities of the reference and those of the stuff from which it was derived.[10] Improving the commutability of physical standards is a perennial preoccupation of reference technicians: they aspire to produce materials that, while suitable for analytical handling, resemble as much as possible the natural state of the matter from which they originated. The ideal is to produce reference materials 'in their natural form',[11] adapted to what a member of staff calls 'a real road sample'.

There are important regulatory reasons for a shift towards increasingly 'realistic' reference materials. Official substances in closer agreement with objects as they can be found in the world approximate the law to the realities of mundane matter, bringing the standard to bear more directly, with fewer transitions and translations, on the stuff of the world. If the purpose of a reference material is to control lay measurements, the more that material approximates the form in which substances are collected, the narrower the gap through which a lay analyst can escape the power of the standard. A reference material in the form of powdery milk can control the measurement of milk powder, but it has nothing to say about fresh milk, or about the quality of the process by which fresh milk was converted into powder. A more 'realistic' reference would have greater control power – or, rather, it will control a slightly different sequence of actions, closer to the initial act of finding and extracting the stuff from the world.

Nothing expresses the utopian aspirations and artistic nature of reference material work better than this pursuit of 'naturalness' or 'realism'.[12] Producing a 'realistic' reference material is utopian because as soon as a substance enters the analytical process its worldliness is irremediably lost. 'Preparing a sample means contaminating it', notes a former member of IRMM;[13] giving the material a form amenable to manipulation and measurement requires contact with instruments and tools, and thus with

---

10 BIPM, *International Vocabulary of Metrology – Basic and general concepts and associated terms (VIM)* (2008) 5.13.
11 Linsinger et al., op. cit., n. 9, p. 1169.
12 The scare quotes around these words are not my own. In their writing, sometimes even in their speech, technicians will use them to qualify what they mean by these terms. When pushed to define what they mean by 'natural' or 'realistic', they will replace these terms with another adjective, also in scare quotes (in some cases they will speak, for instance, of 'fresh' materials). The point that these terms are relative to a particular standard is, needless to say, painfully obvious to reference technicians.
13 B. Griepink et al., 'Concepts of purity' (1982) 19 *Analytical Proceedings* 405–11, at 406.

other materials, altering forever the original state of the stuff. Every instance of material interaction, every contact of the substance in question with any other material – water or air, metals, bacteria, chemicals, plastics, and so on – each step of laboratory processing will compromise in one way or another the state in which matter was sourced.

The original condition of the material can never be preserved. It can, however, be *reconstituted*. This is why the attempt to produce increasingly commutable references, material standards that in some important way are truer to their worldly form, is fundamentally an artistic pursuit. Naturalness must be recreated, a process that far exceeds in sophistication and skill the mere preservation of form. The goal is not the maintenance of the source material in its unadulterated state but, rather, the production of a particular kind of verisimilitude.

That this is a matter of artistic recreation rather than of mere preservation is made clear by the fact that the more 'naturalistic' the reference material, the more complex the laboratory processes involved in its fabrication. As the Reference Materials Unit proceeds towards more 'realistic' standards, the laboratories get larger, the machineries more expensive, and the paper trails longer and more convoluted. Producing a reference material for 'fresh' fish, say, like BCR-718 (tin cans of moist herring tissue with certified quantities of organic pollutants) is a vastly more complicated endeavour than the fabrication of a matrix of freeze-dried fish (ERM-CE477), or fish oil (BCR-349), previous standards for measuring pollution in the marine environment. The difference is that in the case of fish powder, or fish oil, the goal was to produce a material as stable and homogeneous as possible, whereas in the case of BCR-718 the reference was fabricated so as to have *the right kind of instability*, to resemble the changeable state of natural fish while still being amenable to analysis. In other words, BCR-718 represents not only a standard for 'fresh herring' or 'fresh fish', but also an official version of 'freshness' in fish.

Fabricating 'fresh' fish that remains inert (as far as the quantities of the PCBs listed in its Certificate are concerned), for a sufficient period of time (one year after purchase), and with the right degree of homogeneity across samples, requires a different, more complex assemblage of people, technologies, methods, and supporting materials. The following extract from the BCR-718 Certification Report describes some of the steps of the homogenization process, and conveys the technical complexity of the effort:

> The complete volume of herring was minced using a mincer (Finis Machinefabriek, Ulft, the Netherlands) in combination with a Fryma mill equipped with toothed rotary knives (Fryma Maschinen AG, Rheinfelden, Switzerland) to a final size of 3.5 mm$^2$. Subsequently, ten batches of ca. 25 kg sample were homogenized for three minutes, after adding 0.02% butylhydroxytoluene (BHT), in a Stephan cutter (Stephan Machines, Almelo, The Netherlands), type UMM/SK25 (made in 1979). After homogenisation each of the ten batches of herring mince (3.5 times 75 kg) was equally divided over seventeen trays, resulting in ten layers per tray. The trays were covered with aluminium

foil and frozen in a blast freezer. The frozen trays were stored in a freezer at −25°C. Before canning the individual trays of herring were homogenized again in a cutter for 3 minutes.

Two features stand out in this description: the care in describing the machinery used for processing the material (including the name of the tool manufacturer, even the year of fabrication), and the addition of extraneous materials – in this particular case butylhydroxytoluene (BHT), introduced to limit the oxidation of the fish muscle, but also the aluminium of the foil, and other materials that help limit the interaction of the herring with the elements – as a means of allowing the reference to stay in a form resembling its natural state for a much longer period that it would been natural for it to do so.

An increasing degree of naturalistic verisimilitude in the reference material typically requires changes in the practices of its users. A more unstable material requires greater care. The instructions for use in the case of 'fish oil' (BCR-349), for instance, were relatively simple: make sure that the ampoules are heated to 40°C for ten minutes before opening, to increase homogeneity. The instructions for the user of the 'fresh' herring are slightly more complicated:

> The opening of the tin can should happen carefully so that no matrix or moisture is spilled. The entire content should be homogenized using a blender and a sub-sample should be taken immediately after homogenizing. The remainder of the sample may be kept in a glass jar at −20°C for later use, but it should be emphasized that the certified values are only guaranteed for the Certified Reference Material kept in the sealed tin can.

In some cases, realism requires a transformation in the source material itself. Worldly specimens with inborn concentrations of the compound of interest, for instance, are preferable to samples altered upon collection. Pigs grown on a diet rich in pesticide, say, provide a more 'realistic' reference material for pesticide in pork fat than fat from ordinary pigs spiked with pesticide after slaughter.

In other cases, the search for ever-more 'naturalistic' references leads to the development of new artificial substances. For example, the production of references for the measurement of organic pollutants in water, a mandate of the EU Water Framework Directive, poses significant challenges for reference technicians: organic compounds do not dissolve in water and, left to their own devices, will float or attach themselves to the walls of the container. 'It is very difficult to make a water reference material where the water actually contains these organic substances in a way that they are also present in rivers or in ground water', a member of staff notes.

IRMM technicians attempt to solve this problem by producing 'humic substances'. Humic substances are not 'water', but they are common in water, and allow the compounds of interest to bind in a stable fashion. IRMM is currently trying to produce a more realistic version of 'water with organic pollutants' by providing its clients with a small amount of humic substance, along with a solution containing a defined amount of the chemical in question. The client is then expected to mix the two in water and produce a material in which the organic substances are dissolved in a more naturalistic fashion. 'It's a bit complicated to use', a reference technician points out. 'It would be much better if we could give them a bottle that they could open and use as such. But it is better than having no reference material at all. It's a step, but it is still far away from being a real water.'

How far can naturalization go? Reference technicians will readily accept that '[a] completely realistic matrix does not exist',[14] that as soon as the worldly stuff is collected, 'the natural state of the material is profoundly transformed, which automatically leads to some form of matrix mismatch'.[15] Referentiality demands transformations in the original material. And yet, reference technicians will keep striving towards ever-greater commutability. In their work, naturalism is a matter of choosing a certain idiom of realism, and the corresponding set of technical practices, rather than faithfully maintaining the accuracy of representation. It demands an ever-more painstaking attention to fabrication, rather than a mere process of preservation. This is why the search for naturalism is virtually endless: it is an attempt to transform materials into what they already are.[16]

## THE REDEMPTION OF MATERIALS

It is plain that ensuring the referentiality of reference materials is a tricky affair. But what about their materiality? In what sense are these artefacts material, and how does that materiality matter? The answer to this question is less straightforward than it might seem. There are, within the reference materials community, different views on the role of materials in the preservation of standards – contrasting interpretations of the relationship between authority and matter.

There is a tradition – it can perhaps be more accurately described as a regulatory aspiration – eager to keep alive the ideal of a progressive dematerialization of standards. The authority of a reference material, this view

---

14 id.
15 F. Ulberth and H. Emons, 'Reference materials: are they fit-for-purpose for food analysis?' (2005) 381 *Analytical and Bioanalytical Chemistry* 99–101.
16 The relevant analogies can be found in the history of art, in the technical practices of different schools of realism. In the same way that Impressionist painters used a patiently applied layering of paint to convey speed of execution and 'uniqueness of the empirical array' (R.E. Krauss, *The Originality of the Avante-Garde and Other Modernist Myths* (1986) 167), reference technicians carry out carefully planned manipulations to fabricate different varieties of naturalistic verisimilitude – whether it is the freshness of a fish, or the solubility of chemicals in water. Sculpture is probably an even more relevant comparison. As in, for instance, the steel works with which Richard Serra represents the natural strength, elasticity, and decay (oxidation) of steel.

goes, should be clearly differentiated from the authority of any given substance. The authority in question pertains to the *value* of a particular measurement; the measurand is always a *quantity* subject to measurement, not the material itself. Matter merely 'carries' or 'embodies' a particularly authoritative measurement result.[17]

Paul De Bièvre, a member of the original Central Bureau for Nuclear Measurements and for many years the editor of *Accreditation and Quality Control*, offers a clear articulation of this view: 'the long – and still persisting – tradition of focusing on the *material*' he argues, 'should be reoriented towards focusing on the *value carried* by the material'.[18] The physical substrate of the reference material – the stuff inside the ampoules and vials – has a merely instrumental function: it is used to transport the value of an original and especially trustworthy measurement, to convey a result that must be preserved. The material is a *vector*, so to speak, but what really matters is what it transports, and that is always a quantity: a certain amount of X, a particular mass of Y. If a reference material is indeed 'truth in a bottle', it would be silly – this view goes – to concentrate too much on the bottle at the expense of the truth.

De Bièvre's critique of the 'material-centered thinking' of many in the reference materials community[19] does not represent a position in a controversy – nobody will contest the correctness of the principle – but rather a call to keep the metrological ideals of the community firmly in focus, to preserve the vision of a progressive de-materialization of standards. Why? Because substances are by definition too unstable to anchor immutable and timeless measurements. When values are incarnated in material artefacts, they 'tend to lose their accuracy by too much contact with the profane world which they calibrate'.[20] This is why reference technicians should strive to produce standards whose authority is 'traceable to values firmly anchored in nature',[21] that is, 'to become independent of man-made artefacts and to anchor measurement results as far as possible in inalterable properties of nature and in units (with their values) derived from these'.[22]

---

17 'Carry' is the verb most commonly used by proponents of this view to describe the interaction between substance and value. 'Embody' is used in BIPM's 2008 edition of the *International Vocabulary of Metrology*: a reference material 'comprises materials embodying quantities as well as nominal properties' (BIPM, op. cit., n. 10, p. 50). De Bièvre speaks also of values 'hidden' or 'buried' in reference materials: P. De Bièvre 'The key elements of traceability in chemical measurement: agreed or still under debate?' (2000) 5 *Accreditation and Quality Assurance* 423–8.
18 De Bièvre, id., p. 425.
19 P. De Bièvre, 'Can unstable materials fulfill a stable "reference" function?' (2002) 7 *Accreditation and Quality Assurance* 515.
20 J. O'Connell, 'Metrology: The Creation of Universality by the Circulation of Particulars' (1993) 23 *Social Studies of Sci.* 129–71, at 151.
21 P. De Bièvre 'Traceable to values firmly anchored in nature' (1999) 4 *Accreditation and Quality Assurance* 323
22 De Bievre, op. cit., n. 17, p. 428.

This is the spirit of the International System of Units (SI), and the most famous example of de-materialization of a standard is the twentieth-century redefinition of 'metre', from the length of an ingot of platinum-iridium kept in a Paris suburb, to the path traveled by light in a vacuum in 1/299,792,458 of a second. The artefactual 'metre' was always in flux – moving it, cleaning it, measuring it, resulted in small but critical changes in its length – whereas light would always travel an identical length, given 1/299,792,458 of a second and a space suitably empty of matter.

It would be inaccurate to describe this view as anti-materialist: try to measure the path traveled by light without material implements, or to produce a perfect vacuum – in other words, try to deliver truth without a bottle. What this position upholds is the moral value of standards that are as independent of particular pieces of matter as possible. Artefacts and substances are too mutable to guarantee stability of reference. Even if a physical incarnation of measurement is unavoidable, the community should strive towards non-material form of objectivity.

Why these reminders of the perils of artefactual objectivity? In his study of the development of intrinsic standards, O'Connell speaks of a 'Calvinist reformation in metrology'.[23] With the substitution of artefact standards by standards produced by physics experiments, contact with high authority was seemingly available to any individual, without the interference of mediating objects and institutions:

> Noticeably absent from the metrology of intrinsic standards is the periodic sacramental redemption from error that equally mark the Catholic theology and the metrology of artefact standards. Instead, the judgment of whether the intrinsic standard is good or bad occurs only once, when it is built, and there is no recognition or provision that correction or comparison, or contact of any sort with the higher authority, will be needed again.[24]

'Periodic sacramental redemption' is a good way of describing the work of reference technicians. Their job, let us remember, is not simply to source original matter and manipulate it to produce an official substance, but also to watch for deviations and degradations, and, every so often, to correct the original assessment or produce a new, immaculate version of the standard. This continuous vigilance rests, as we saw, on a number of organizational practices, including constant textual description of their activities, and a set of ingrained routines for the handling of stuff. 'Preservation, as it is understood by IRMM,' write members of the Institute, 'consists of two aspects: firstly, it comprises all efforts to prevent degradation; secondly, it consists of the measures taken to detect degradation for those changes where even the most cautious prevention failed.'[25]

---

23 O'Connell, op. cit., n. 20, p. 154.
24 id.
25 Linsinger et al., op. cit., n. 9, p. 1169.

In the everyday life of reference technicians, organizational stability, practical routines, and technical dexterity are much more significant sources of authority for the certified materials than any direct access to Nature's values. Their overarching concern is a certain kind of naturalism in representation – the preservation of a certain set of qualities in a suitably inert form – rather than ensuring a connection with values anchored in Nature. That Nature (with capital 'N') is appealing to metrological idealists because it is presumed inalterable, whereas the nature that reference technicians have to contend with is visibly mutable and volatile.

The debate over the materiality of reference materials reflects this tension between a materialist 'nature' of constant degradation and an idealist 'Nature' of infinite permanence. This tension is constitutive of what Mody and Lynch have described as 'test objects', material things that are used to discipline an analytical system and integrate a community of observers. Like model organisms, cell lines, type specimens, and prototypes, reference materials possess 'an ontological status that confounds the familiar distinction between conditions of observation and objects of reference' – they 'are not readily compartmentalized as natural objects, engineered things, representation devices or ready-to-hand instruments'.[26] This fluidity – enabling different dimensions of the object to emerge in different moments of use – is critical, given the many functions reference materials are called to perform: instantiation of a legal category, tool for the calibration of analytical machinery, and consensus-forming artefact for the community.[27]

## DISCUSSION: THE TRANSITIONAL OBJECTS OF THE LAW

The world seen from the vantage point of the IRMM is one in continuous change, made up of materials that are exquisitely sensitive to their environment and thus in constant transition. There is no better way of learning to appreciate the liveliness of our material world, the vibrancy of matter,[28] than spending some time with scientists and technicians so obstinately dedicated to rendering it inert. The staff of the IRMM can attest to the fact

---

26 C. Mody and M. Lynch, 'Test objects and other epistemic things: a history of a nanoscale object' (2010) 43 *Brit. J. for the History of Science* 423–58, at 426.
27 We can say of the work of the reference technicians what Daston says of botanists and their ability to use a type specimen to instantiate a plant species: that theirs is 'a radical solution to the several problems of how to compress the many into one, to render the abstract via the concrete, and to tether words to things and hence akin to the dilemmas of political representation, literary personification (or, for that matter, theological incarnation), and linguistic reference' (L. Daston, 'Type Specimens and Scientific Memory' (2004) 31 *Critical Inquiry* 153–82, at 157).
28 J. Bennett, *Vibrant Matter: a political ecology of things* (2009).

that, as Ingold notes, 'materials may lie low but are never entirely subdued'.[29]

Materials are dynamic, stubbornly fluid, and difficult to pin down. Producing static versions of them – even if it is only temporarily, in relation to specific parameters, and within a certain range of uncertainty – is an inordinate task. In the world of reference technicians, the self-identical nature of substances can never be taken for granted. Selfsameness is constantly undermined by the intimacy and interaction between different materials, by what Fox-Keller describes as 'the capacities of material entities to shape, inform and effect other material entities with which they come into contact'.[30] The elements conspire against the stability of physical standards, transforming matter in ways that compromise its referential value. This is a world in continuous 'degradation' or 'decay', a process reference technicians strive to interrupt, if only provisionally, by fabricating unusually homogeneous and stable physical versions of worldly stuff. In their pursuit of inertness, they resemble other experts in the art of arresting the transformation of matter: art conservators,[31] natural history curators,[32] botanists,[33] or taxidermists.[34]

It is striking how our customary distinctions for classifying matter – organic/inorganic, physical/chemical, living/non-living – lose their salience in the reference materials laboratory. Reference technicians treat *all* materials as if they were living matter. They are constantly attuned to the plasticity of substances and the temporality of any of their forms.[35] The perspective of the reference technician upends those models of reality in which 'materiality' is synonymous with 'solidity', 'stability', 'regularity' or 'permanence'. The suspicion of the metrological idealist is here enlightening: nothing material stays the same long enough to guarantee an unalterable social order.[36]

Achieving a relative inertness of matter is a means towards an ever greater and more astonishing goal. The true triumph of the reference material is to render law mundane – to mark physical objects with the categories of the

---

29 T. Ingold, 'Materials against materiality' (2007) 14 *Archaeological Dialogues* 1–16, at 10.
30 E. Fox Keller 'Towards a science of informed matter' (2011) 42 *Studies in History and Philosophy of Biological and Biomedical Sciences* 174–9, at 177.
31 B. Brown, 'Objects, Others, and Us (The Refabrication of Things)' (2010) 36 *Critical Inquiry* 183–217.
32 M. Rossi, 'Fabricating Authenticity: Modeling a Whale at the American Museum of Natural History, 1906–1974' (2010) 101 *Isis* 338–61.
33 Daston, op. cit., n. 27.
34 D. Haraway, 'Teddy Bear Patriarchy: Taxidermy in the Garden of Eden, New York City, 1908–36' (1984–85) 11 *Social Text* 19–64.
35 H. Landecker, *Culturing Life: How Cells Became Technologies* (2007).
36 N. Marres and J. Lezaun, 'Materials and Devices of the Public: An Introduction' (2011) 40 *Economy and Society* (forthcoming).

law. The workers at IRMM are the indispensable technicians of the law, yet their labours remain inconspicuous.[37] The effacement of the labour involved in making legal categories material is evident in the lack of attention of legal scholars to the fabrication of the artefacts that incarnate legal entities. Even when the focus is on 'standardization', or 'harmonization', the emphasis tends to be on the production of texts, rules, and criteria – in other words, on processes of writing that continue and resemble the textual fabrication of the law. The sociological study of commensuration practices, on the other hand, while concerned with 'the transformation of different qualities into a common metric',[38] has little to say about material commutability, preferring to focus on instances where quantities and numerical values provide the basis of comparison. When we visit IRMM, however, what we observe is not only a great deal of writing and calculating, but also a multitude of handling practices, machineries, and physical labours, all in the service of giving law a material form, of hardening the law so that it can survive contact with the world.

Describing the production of scientific reference, Latour notes that 'the sciences do not speak of the world but, rather, construct representations that seem always to push it away, but also to bring it closer'.[39] The law does not speak of the world either, unless worldly stuff can be classified on the basis of legal criteria. In some cases, that requires a material intermediary, a sort of transitional object. With a reference material, the stuff of the world approximates a legal ideal by adopting a decidedly unnatural (but often 'realistic') inertness, while the legal category in question has to contend with the continuous instability and fluidity of matter. The pragmatic sanction of materials is thus never a matter of *applying* a legal principle to a singular object, of fitting the abstract ideal to the mundane exemplar. Nor is it a mere attempt to *embed* a value – legal or otherwise – in a piece of matter. It is, rather, the manufacture of *radically original legal substances*, substances that allow the law to become of the world.

---

37 Historical studies of metrology have emphasized the elision and invisibility of the practical labour that accompanies the production of stable references. 'The immense labour required to set up such unalterable standards', Simon Schaffer writes of Victorian metrology, 'was always accompanied by a deliberate effort to efface this labour' (S. Schaffer, 'Accurate Measurement is an English Science' in *The Values of Precision*, ed. M. Norton Wise (1995) 135–72, at 136).
38 W.N. Espeland and M.L Stevens, 'Commensuration as a social process' (1989) *Ann. Rev. of Sociology* 313–43, at 314.
39 B. Latour, *Pandora's Hope: Essays on the Reality of Science Studies* (1999) 30.

# The Regulation of Nicotine in the United Kingdom: How Nicotine Gum Came to Be a Medicine, but Not a Drug

Catriona Rooke,* Emilie Cloatre,** and Robert Dingwall***

*This article explores the utility of actor-network theory (ANT) as a tool for socio-legal research. ANT is deployed in a study of the evolution of divided regulatory responsibility for tobacco and medicinal nicotine (MN) products in the United Kingdom, with a particular focus on how the latter came to be regulated as a medicine. We examine the regulatory decisions taken in the United Kingdom in respect of the first MN product: a nicotine-containing gum developed in Sweden, which became available in the United Kingdom in 1980 as a prescription-only medicine under the Medicines Act 1968. We propose that utilizing ANT to explore the development of nicotine gum and the regulatory decisions taken about it places these decisions into the wider context of ideas about tobacco control and addiction, and helps us to understand better how different material actors acted in different networks, leading to very different systems of regulation.*

---

* Centre for Population Health Sciences, University of Edinburgh, Medical School, Teviot Place, Edinburgh EH8 9AG, Scotland
catriona.rooke@ed.ac.uk
** Kent Law School, University of Kent, Canterbury, Kent CT2 7NX, England
e.cloatre@kent.ac.uk
*** Dingwall Enterprises/Nottingham Trent University, 109 Bramcote Lane, Wollaton, Nottingham NG8 2NJ, England
robert.dingwall@ntu.ac.uk

Funding from the ESRC and ASH is gratefully acknowledged. Particular thanks are due to Ann McNeill for valuable feedback on various drafts of this paper and, with Deborah Arnott, for invaluable contributions to the project, and to the informants who generously gave their time to this research.

## INTRODUCTION

This article takes as a starting point a regulatory problem highlighted by the public health community. The issue was outlined in a *Lancet* 'viewpoint' piece:

> ... the most dangerous and addictive nicotine products [smoked tobacco] remain only slightly regulated, in great disproportion to their hazard, and are freely available and widely used ... By contrast, medicinal nicotine products, which are the safest source of nicotine, are generally subject to the highest levels of regulation since they are generally classified as drugs. This is almost certainly a major disincentive to new product development and innovation, and to market competition to create better and more effective cigarette substitutes.[1]

Tobacco and medicinal nicotine (MN) products (nicotine gums, patches, and so on) are regulated under different frameworks in the United Kingdom. Tobacco is governed by a variety of different legal instruments which control the way that products can be advertised, marketed, sold, and consumed but exert little control over product content. MN products are regulated by the Medicines and Healthcare products Regulatory Agency under the Medicines Act 1968. Manufacturers of MN must provide evidence to satisfy the regulators of products' safety and efficacy, and give consumers extensive product information, including restrictions and cautions about their use, in contrast to the simple warnings on tobacco packaging.

We explore the evolution of this divided regulatory responsibility by tracing the development and licensing of nicotine gum, the first MN product. Through this account we examine the utility of actor-network theory (ANT) ideas for socio-legal research. We propose that utilizing ANT to analyse the regulatory decisions taken about nicotine gum in the United Kingdom will place these decisions into the wider context of ideas about tobacco control and addiction, and help us to understand better how different nicotine products came to be enrolled into, and consequently shaped by, different systems of regulation.

Much has been written on the history of tobacco.[2] In particular, the discovery of the link between smoking and lung cancer and the subsequent policy response has been comprehensively investigated. Berridge notes that accounts of the last half century have been dominated by 'activist' histories using tobacco industry documents, and developments in United States policy.[3]

---

1 J. Britton and R. Edwards, 'Tobacco smoking, harm reduction, and nicotine product regulation' (2008) 371 *Lancet* 441–5, at 441.
2 See V. Berridge, *Marketing Health: Smoking and the Discourse of Public Health in Britain, 1945–2000* (2007); A.M. Brandt, *The Cigarette Century* (2007); J. Goodman, *Tobacco in History: The Cultures of Dependence* (1993); S. Lock et al., *Ashes to Ashes: The History of Smoking and Health* (1998); S. Wagner, *Cigarette Country* (1971).
3 Berridge, id.

Fewer accounts have focused on the United Kingdom, and there is especially little linking the development of MN into this history. Using the idea of policy networks, Read explored the relationship between the British government and the tobacco industry.[4] Berridge's *Marketing Health* is a detailed account of British tobacco policy, which uses tobacco as a lens through which to examine the 'stages of change' in public health during the latter half of the twentieth century.[5] Berridge's account of the period during which nicotine gum was developed highlights the medicalization of smoking. She draws attention to the different networks within which ideas about smoking were embedded and the changing positions of tobacco and nicotine.

This article utilizes ANT to build on Berridge's account whilst focusing more specifically on the period during which nicotine gum was developed to explore the origins of this divided regulatory responsibility. Firstly, we discuss key ANT concepts, outlining how they are used here and the ways in which they build on previous accounts. We then describe how the study was carried out and the sources drawn on. A brief outline of the relevant moments in the history of tobacco use and control is provided, before narrowing the focus to explore events surrounding the development of nicotine gum and its emergence on the market as *Nicorette*.

## ACTOR-NETWORK THEORY

ANT emerged in the 1980s from science and technology studies (STS) and is particularly associated with the work of Callon,[6] Latour,[7] and Law.[8] It cannot, however, be described as a unified approach and has continued to evolve with subsequent studies developing ANT ideas in various different directions.[9] Any account of this large and shifting body of ideas is necessarily partial; therefore we focus on elucidating those concepts central to our account.

---

4 M.D. Read, *The Politics of Tobacco: Policy Networks and the Cigarette Industry* (1996).
5 Berridge, op. cit., n. 2.
6 M. Callon, 'Some Elements of a Sociology of Translation: Domestication of the Scallops and the Fishermen of St Brieuc Bay' in *Power, Action and Belief: A New Sociology of Knowledge?*, ed. J. Law (1986) 196–233.
7 B. Latour, *Science in Action* (1987).
8 J. Law, 'Technology and Heterogeneous Engineering: the Case of Portuguese Expansion' in *The Social Construction of Technological Systems*, ed. W.E. Bjiker (1987) 111.
9 For example, A. Mol, 'Missing links, making links: the performance of some atheroscleroses' in *Differences in Medicine: Unravelling Practices, Techniques and Bodies*, eds. M. Berg and A. Mol (1998) 144–65; V. Singleton and M. Michael, 'Actor-Networks and Ambivalence: General Practitioners in the UK Cervical Screening Programme' (1993) 23 *Social Studies of Sci.* 227–64.

Latour suggests that networks be seen as 'the summing up of interactions through various kinds of devices, inscriptions, forms and formulae, into a very local, very practical, very tiny locus'.[10] The network metaphor in ANT is used as a means of focusing attention on the relationships between actors, and on the negotiations involved in forming and maintaining connections, and subsequently actor-networks. This highlights the ongoing work needed to overcome resistances and to preserve fragile associations: a network that appears durable is an achievement.[11] The concept of translation describes the 'dynamic process through which facts, concepts, and physical entities move from site to site and are either reinforced and solidified or else contradicted or undermined.'[12] Translation is the process by which entities enrol and order each other and come to speak for and configure other entities.

ANT emphasizes that actor-networks are *heterogeneous*.[13] Actor-networks can be assembled out of human, material, and discursive elements. Another quality of actor-networks is that they often become simplified – *punctualized* or *black-boxed* – so that complex webs of relations come to appear as single entities. They may also contain *obligatory passage points* where one actor comes to shape the network, controlling the enrolling and ordering of other actors.

A final idea to note is that of *symmetry* in explanation, which underlines the need to avoid a priori assumptions in approaching the situation being studied.[14] For ANT, entities are constituted through relations and interactions as 'a continuously generated effect of the webs of relations within which they are located'.[15] Categories such as social and legal, human and non-human are seen as the outcomes of assemblages of heterogeneous elements. Much of the criticism of ANT has revolved around its conception of symmetry and use of the network metaphor. Lee and Brown warn that ANT's inclusion of non-humans leaves nothing outside – no other – and risks the production of another grand narrative.[16] Star questions the political consequences of ANT's processes of delegation, raising the issue of how some human perspectives win over others and suggesting that a network looks different depending on where you stand in relation to it.[17] Strathern

---

10  B. Latour, 'On recalling ANT' in *Actor Network Theory and after*, eds. J. Law and J. Hassard (1999) 15–25, at 17.
11  Law, op. cit., n. 8.
12  M. Valverde et al., 'Legal knowledges of risk' in *Law and Risk*, ed. Law Commission of Canada (2005) 86–120, at 86.
13  Law, op. cit., n. 8.
14  Callon. op. cit., n. 6.
15  J. Law, 'Actor Network Theory and Material Semiotics' in *The New Blackwell Companion To Social Theory*, ed. B.S. Turner (2008) 141–58.
16  N. Lee and S. Brown, 'Otherness and the Actor Network: The Undiscovered Continent' (1994) 37 *Am. Behavioral Scientist* 772–90.
17  S.L. Star, 'Power, technologies and the phenomenology of conventions: on being allergic to onions' in *A Sociology of Monsters: Essays on Power, Technology and Domination*, ed. J. Law (1991) 26–56.

suggests that the network metaphor has 'properties of autolimitlessness; that it is a concept which works indigenously as a metaphor for the endless extension and intermeshing of phenomena', raising the problem that the networks under study may be, theoretically, without limit.[18]

Whilst STS as a whole has engaged with law,[19] ANT accounts have generally been rather silent on the subject of the legal.[20] Socio-legal scholars have, however, started to consider the utility of ANT.[21] In a study examining the impact of TRIPS and pharmaceutical patents on health in Djibouti, Cloatre uses the concept of *socio-legal objects,* defined as 'objects with a legal origin/dimension studied in their social action through networks and connections', to investigate the links between written legal rules and their social actions.[22] She views patented drugs as a 'particular type of hybrid, made up both from the complexity of pharmaceutical patents and of drugs'.[23] Understanding pharmaceutical regulation not as written prescriptions but as socio-legal objects leads to questions about how regulations act and what effects they generate. The material and its link to regulation are conceived of in a more interdependent way where law is both the result of socio-technical assemblages and becomes part of specific materials, so the 'things' followed during analysis are themselves shaped and defined by the legal and regulatory frameworks that they carry.

ANT builds on accounts such as those by Read[24] and Berridge[25] by moving beyond policy networks as stable sets of connections to look at the shifting heterogeneous assemblages that constitute them. It directs us to examine how these networks come to appear as stable. The concepts of translation and heterogeneity urge the analyst to be attentive to the circulation of non-human actors and the translations they undergo, the actors they enrol and the heterogeneous connections from which they themselves are constituted.

To do ANT, Latour advises that one must 'follow the actors' and trace the connections that they construct.[26] To trace the associations made by MN

---

18 M. Strathern, 'Cutting the network' (1996) 2 *J. of the Royal Anthropological Institute* 522-35, at 522.
19 See S. Jasanoff, 'Making order: law and science in action' in *The Handbook of Science and Technology Studies*, eds. E.J. Hackett et al. (2008, 3rd edn.) 761–86.
20 Latour has recently turned his attention to law and its activities: B. Latour, *The Making of Law: An Ethnography of the Conseil D'Etat* (2009).
21 E. Cloatre, 'Trips and Pharmaceutical Patents in Djibouti: An ANT Analysis of Socio-Legal Objects' (2008) 17 *Social & Legal Studies* 263–81; D. Cowan and H. Carr, 'Actor-network Theory, Implementation and the Private Landlord' (2008) 35 *J. of Law and Society* 149–66; Valverde et al., op. cit., n. 12.
22 Cloatre, id., p. 264.
23 id., p. 273.
24 Read, op. cit., n. 4.
25 Berridge, op. cit., n. 2.
26 B. Latour, *Reassembling the Social: An Introduction to Actor-Network-Theory* (2005).

and cigarettes over time we have drawn on a range of primary, secondary, documentary, and interview sources. In discussing tobacco use and control more broadly we draw on various secondary historical accounts.[27] Focusing on the period during which nicotine gum was developed, there are interviews with key actors (Michael Russell and Ove Fernö) in the public domain.[28] We utilize material from the ASH archive at the Wellcome Library and the Ministry of Health papers at the National Archive, as well as scientific papers and reports published at the time. To supplement this material, we draw on semi-structured, qualitative interviews conducted with four informants who were involved with the development of nicotine gum. Topic guides included questions on how the informant came to work in the field of tobacco control, the process of developing and marketing nicotine gum, the development of ideas about the role of nicotine in smoking, and changes in the field, and then moved towards current issues in the regulation of nicotine and the development of new MN products. The next section briefly recounts the key historical transformations that tobacco has undergone.

## TRANSFORMATIONS IN TOBACCO USE

The tobacco plant is indigenous to the Americas. It was 'discovered' by Europeans in the fifteenth century and its use quickly spread throughout much of the world. Until the nineteenth century, tobacco use in Europe took a variety of forms, both smoked and smokeless. During the nineteenth and first half of the twentieth century, various developments in tobacco production and cigarette manufacture, along with parallel shifts in manufacturing, advertising, and promotion came together to constitute the modern cigarette.[29] With this came large increases in tobacco consumption – predominantly in the form of cigarettes – and cigarette smoking became increasingly accepted and normalized.

Early in the twentieth century a parallel increase in cancer of the lungs was noticed by vital statisticians.[30] This new connection sparked the modern period of investigation of smoking and health, during which the cigarette was enrolled into a new network of medical practices and techniques. The cigarette was tentatively linked with a disease, lung cancer, and began to be translated into something risky. During the 1950s several epidemiological studies demonstrating increased risk of lung cancer in smokers were published. In 1962 the Royal College of Physicians (RCP) published a report on

---

27 In particular, Berridge, op. cit., n. 2; Goodman, op. cit., n. 2.
28 O. Fernö, 'Conversation with Ove Fernö' (1994) 89 *Addiction* 1215–26; M.A.H. Russell, 'Conversation with Michael A.H. Russell' (2004) 99 *Addiction* 9–19.
29 See Goodman, op. cit., n. 2.
30 Berridge, op. cit., n. 2; Brandt, op. cit., n. 2.

*Smoking and Health*.[31] The report gave legitimacy to the earlier studies and their methods, and brought the evidence into the public and policy domain. It stabilized the cigarette as a risky object within medical networks, which carried an embedded notion of risk when travelling to new sites. However, despite growing scientific consensus, there was still unwillingness in government to take an active role in intervening in what was considered to be an individual habit.

By the 1970s the cigarette-lung cancer connection had further stabilized; cigarettes had been shaped into a sufficiently risky actor to push the government to begin taking a more active role in tobacco control. In the 1970s and 1980s, new actors were enrolled into the tobacco network and important discussions took place on why people smoked and what were effective and acceptable ways to reduce the damage caused by smoking; on the role of nicotine in the 'smoking habit'; and how an already established substance, tobacco, and a new and unique substance, nicotine gum, should be controlled. Our account of this time examines how these questions were answered and by whom, by following the translations that occurred as nicotine gum and the connected idea of nicotine addiction were constructed and extended to various sites.

## REGULATING NICOTINE

The active principle of tobacco was discovered in 1809 by a French chemist and named nicotine. In the early 1970s Michael Russell, a psychiatric researcher at the Addiction Research Unit (ARU) in London, reviewed the evidence on the role of nicotine in smoking.[32] A number of studies had previously examined the actions of nicotine or smoking on the body, showing effects on the peripheral and central nervous system. Others took a different approach and investigated the relationship between smoking patterns and personality. Russell was interested in understanding the 'many and various' actions of nicotine in the 'smoking habit', rather than the attributes of smokers.[33] His view was that nicotine is the key reason people smoke: 'If it were not for the nicotine in tobacco smoke people would be little more inclined to smoke cigarettes than they are to blow bubbles or light sparklers.'[34] A colleague remembered:

> Back in 1970 he decided that smoking was really all about giving yourself nicotine, which was a very unorthodox view at that time because people were generally unaware of pharmacological factors.[35]

31 Royal College of Physicians (RCP), *Smoking and Health* (1962).
32 M.A.H. Russell, 'Cigarette smoking: natural history of a dependence disorder' (1971) 44 *Brit. J. of Medical Psychology* 1–16.
33 id.
34 id., p. 7.
35 Interview six, clinical psychologist.

Russell began working to strengthen and expand on this link. He was concerned with why people start, continue, and stop smoking, and what it is about nicotine that motivates people to smoke. Russell's work also demonstrated a desire to tie his insights about nicotine to smoking policy, for example, the focus of the team at the ARU on nicotine led them to consider the ratio of tar to nicotine in cigarettes. For Russell, unlike tar and carbon monoxide (CO), there was 'no evidence that nicotine is harmful in smoking doses' and no doubt that nicotine was the 'primary addictive component of tobacco'.[36] He was clear that the problem is the harm smoking causes, not that it is addictive. This led him to advocate 'safer smoking' approaches such as 'low-tar, medium-nicotine' cigarettes.[37]

Nicotine was also central in the story of nicotine gum. According to Ove Fernö, research director at Leo Pharmaceutical Company in Sweden, this story started in 1967 with a letter from a friend suggesting:

> ... a tobacco substitute for oral use in such a way that suitable doses of nicotine could be administered, which would prevent the user from being exposed to the many harmful constituents of tobacco smoke ... he had noticed that submariners, because they were not allowed to smoke, could switch to chewing tobacco in the boat without too much difficulty.[38]

Like Russell, Fernö started by constructing nicotine as an obligatory passage point in developing a substitute for smoking. He stated that he did not have any doubt that 'nicotine was the main element in the smoking habit'.[39] As research director, Fernö had the resources and freedom to act on his definition. Drawing on his understanding of the absorption of nicotine from tobacco products and various trials, he settled on a nicotine gum as the appropriate form for his cigarette substitute. A belief in the importance of the role of nicotine in smoking led some researchers in Sweden to begin working with nicotine gum. One recounted:

> I had a very open and positive expectation ... [of the gum]. At that time I was fairly convinced that this is not just a habit; there is an element of drug addiction here and nicotine is probably the culprit.[40]

Having first become interested in nicotine gum at a conference in 1975, and afterwards visiting Fernö in Sweden,[41] Russell, who by this point was heading a tobacco group within the ARU, began conducting various trials with the gum. These highlighted the way that smokers regulate their levels of nicotine and reinforced that 'nicotine is an important determinant of

---

36 M.A.H. Russell, 'Low-tar medium-nicotine cigarettes: a new approach to safer smoking' (1976) 1 *Brit. Medical J.* 1432–3.
37 M.A.H. Russell, 'Smoking problems: an overview' in *Research on Smoking Behavior: NIDA Research Monograph 17*, eds. M.E. Jarvik et al. (1977) 13–34.
38 Fernö, op. cit., n. 28, p. 1216.
39 id.
40 Interview two, psychologist.
41 Russell, op. cit., n. 28.

smoking'.[42] For the network being put together by Russell and Fernö, the trials demonstrated the role of nicotine in generating a disease, nicotine addiction, and the role of nicotine gum in treating it.

Enrolling Fernö's company into this programme of action, however, required work. Russell commented that 'the company president and the scientific advisory committee had resisted supporting the gum for some 4–5 years'.[43] A Swedish psychologist who worked with the gum shed light on this:

> The pharmaceutical company was mainly involved with anti-cancer drugs and antibiotics, painkillers ... they said ... what is the indication? Smoking cessation; that's not a disease. So, what's the substance? Nicotine; that's a poison. What's the administration form? A chewing gum; oh, that's candy ... nothing about it really fitted into being a scientific, sophisticated product for a research-orientated pharmaceutical company.[44]

Moreover he explained that 'nicotine was a first class poison ... and it had other sorts of baggage around it'. Fernö also noted that 'chewing gum is not a typical product for a pharmaceutical company. Most people in the company did not realize the potential in this idea.'[45] Nicotine gum did not fit any of the categories available to a pharmaceutical company; many actors defined the smoking problem differently and nicotine addiction was a contested disease classification.[46] Russell and Fernö underlined the importance of studies undertaken by Russell's team, which demonstrated the safety and efficacy of nicotine gum, in enrolling Leo and further legitimating the gum.[47]

Licensing the gum in Sweden was also problematic as no regulatory agency would accept responsibility. Fernö recollected the drug authority's ruling that:

> Agents acting against the desire to smoke were not drugs. The consequence of this decision was that a chewing gum containing nicotine was classified as a foodstuff, as chewing tobacco and snuff had been for a long time. However if such agents could be proved to cure a disease caused by smoking they could be considered drugs.[48]

Believing that the gum would be marketed as a food, the food regulatory body was approached;[49] however:

---

42 M.A.H. Russell et al., 'Effect of nicotine chewing gum on smoking behaviour and as an aid to cigarette withdrawal' (1976) 2 *Brit. Medical J.* 1391–3.
43 Russell, op. cit., n. 28, p. 13.
44 Interview two, psychologist.
45 Fernö, op. cit., n. 28, p. 1221.
46 G.C. Bowker and S.L. Star, *Sorting things out: classification and its consequences* (1999).
47 Fernö, op. cit., n. 28; Russell, op. cit., n. 28.
48 Fernö, id., p. 1223.
49 Interview four, pharmacist at Leo.

The new head [of the Swedish food regulatory body] said of 'course that it is absolutely impossible: you can't add nicotine to food, it's a poison'. So that road was blocked. And the drugs people still said, 'no it's not a disease; you can't apply for a registration'.[50]

When nicotine gum initially encountered Swedish food and medicines regulations, it fitted neither framework: a substance containing nicotine could not be food; however, a treatment without a disease could not be a medicine:

> Drugs are for indications, so diseases. Smoking cessation is not a disease ... it's a description of something so it didn't really fit the medical model in a sense. If it had said treatment for tobacco dependence, yes, that would have been more appropriate. But at that time there was no diagnosis, tobacco dependence: that didn't exist, it came later.[51]

The positioning of nicotine gum within a pharmaceutical company, itself enmeshed in complex webs of medicines regulation, meant that, for the nicotine gum network to be extended, it had to be incorporated into an existing regulatory regime. The regulations acted as an obligatory passage point through which the gum could not pass, being able to act as neither food nor medicine, and it was prevented from entering the Swedish market until it was successfully able to act as a medicine within the drug regulatory network. Within the network being constructed by Russell and Fernö, nicotine gum was a treatment for a disease; however, the attitude of Leo Pharmaceuticals and the struggle to get nicotine gum licensed in Sweden illustrate the challenges in stabilizing and extending this network. To better understand the difficulty in extending the network to the United Kingdom, we now turn our attention to the emergent tobacco control coalition there.

## CONTROLLING CIGARETTES

In 1971 the RCP increased pressure on the government by publishing a second report. In the same year, the government, maintaining a tradition of industry-government negotiation and cooperation,[52] began to address the smoking problem using a soft-law instrument – voluntary agreements with industry. The first such code focused on labelling and advertising. In response to these changes, two new actors were created to deal with smoking: the Independent Scientific Committee on Smoking and Health (ISCSH) and Action on Smoking and Health (ASH). They approached smoking in different ways. ASH was set up by the RCP in response to a perceived need for a central information point to inform the public about the

---

50 id.
51 Interview two, psychologist.
52 Read, op. cit., n. 4.

risks of smoking. They were to act as an external pressure group focused on changing opinion on smoking and putting pressure on government for change.[53] Following the voluntary agreement, the Secretary of State referred the issue of cigarette packet labelling to a new expert committee. The ISCSH evolved from this process to fulfil a perceived need for systematic and impartial scientific advice on smoking and health.

To understand the outlook of the ISCSH, it is worth briefly exploring some of the relationships in which it was enmeshed. Regulatory networks for both licit and illicit drugs were reshaped during this period. Drug safety regulation went through extensive change in the 1960s, following the Thalidomide tragedy which exposed serious shortcomings in the control of drugs, resulting in the passing of the Medicines Act 1968.[54] This replaced most of the previous statutes relating to medicines, poisons, and drugs and established the Medicines Commission and Committee on the Safety of Medicines (CSM) to which all new drugs would be submitted. It controls the manufacture and distribution of medicines – with safety, quality, and efficacy criteria to be met. Berridge underlines the importance of connections between this network and the ISCSH, and that in both areas it was quite normal to retain relationships with industry.[55]

The ISCSH's remit included receiving 'full data about the constituents of cigarettes', reviewing the research into less dangerous smoking, and advising on 'testing the health effects of tobacco and tobacco substitutes'.[56] The ISCSH's definition of the smoking problem began with the cigarette as central, and translated it into a question of how to transform the cigarette to make it a less risky actor. In 1975 the ISCSH reported on the preparation of guidelines for the testing of cigarettes containing tobacco substitutes, and discussed additives in tobacco products and the composition of cigarette smoke.[57] Its second report continued and extended this programme.[58] It outlined progress in the testing of substitutes and the decision to allow them to be marketed with specific conditions. It also moved on to discuss 'lower-risk' cigarettes, with a particular interest in the reduction of tar, nicotine, and CO. The focus was on making small changes to the cigarette actor-network, particularly to tobacco itself, which was unblack-boxed and several of its constituents constructed as risky.

Meanwhile, for the first couple of years the role of ASH remained fluid. Definitions of its function differed and its activities were 'low-key'.[59] Its

53 Berridge, op. cit., n. 2.
54 S. Anderson, *Making Medicines: A history of pharmacy and pharmaceuticals* (2005).
55 Berridge, op. cit., n. 2.
56 Independent Scientific Committee on Smoking and Health (ISCSH), *Tobacco substitutes and additives in tobacco products* (1975) Appendix I.
57 id.
58 ISCSH, *Developments in tobacco products and the possibility of 'lower-risk' cigarette* (1979).
59 Berridge, op. cit., n. 2.

earlier aims encompassed both stopping people smoking and, like the ISCSH, lowering the risk of their smoking.

During this time, the Minister for Health, David Owen, was concerned with the relationship between government and the tobacco industry: 'Until you can negotiate with the industry, with them knowing that you can, that legislation is a realistic possibility, you will never have a proper negotiating machinery.'[60] Believing legislation would be unacceptable, he proposed using the Medicines Act 1968 to control the tobacco substitutes and additives being considered by the ISCSH, with the future intention of also controlling tobacco this way:

> The order ... will ensure that those tobacco products consisting of or containing a substitute for tobacco or an additive to the tobacco would need a product licence from the Government. Such a licence would be granted on advice received from a statutory committee on the safety of the product. This committee would be established under section 4 of the Medicines Act and be based on the existing independent scientific committee.[61]

This would, as for most other licit drugs, have positioned the Medicines Act 1968 as an obligatory passage point in the marketing of tobacco substitutes and increased the government's control over product content. A draft bill was produced and the regulation passed all the legislative scrutiny; however, there were concerns about the order procedure and politicization of the Medicines Act.[62] In 1976 Owen moved to the post of Foreign Secretary. Without him, the initiative lost momentum. Owen later suggested that 'the legislation was never brought forward because the Labour government feared too much the effect on the voters and the capacity of the tobacco industry to generate criticism'.[63] This incident highlights the nature of law making as a fragile, contingent process: as a socio-legal object, a legislative text is the result of the right gathering of connections; if one fails, it may not come into being. Legislation is also essential in shaping the material: if the bill had been passed, cigarettes could have become a different type of socio-legal actor than they are today; regulated differently, they would be quite different.

The ISCSH's approach shifted during the 1970s. After the failure to bring tobacco substitutes under the Medicines Act 1968 and the unsuccessful launch in 1977 of a product containing a tobacco substitute that the ISCSH had worked on with Imperial Tobacco,[64] it focused on making existing cigarettes safer, mainly through the reduction of nicotine and tar. The committee's third report investigated issues in the development of less harmful cigarettes: the role played by tar, nicotine, and CO in smoking and

---

60 W. Norman, Interview with David Owen, 20 January 1976 (Wellcome Library: SA/ASH/r.24).
61 D. Owen MP, 903 *H.C. Debs.* cols. 810–11 (16 January 1976).
62 Interview with David Owen, 2009
63 D. Owen, *Our NHS* (1988) 147–8.
64 Berridge, op. cit., n. 2, pp. 142–6.

their levels in cigarettes; investigations of other components of tobacco smoke; and monitoring of the health effects of modified products.[65] A report published in 1988 represented a continuance of this more cautious programme.[66] Notably, it was the first of the ISCSH's reports to begin by emphasizing that people should be encouraged to stop smoking, before going on to say that, if they are not able to, they should be encouraged to smoke less harmful products. Through the 1970s and 1980s the ISCSH maintained a close relationship with the tobacco industry and tried to translate the smoking problem into a question of how to transform the cigarette to make it a less risky actor.

Whilst the ISCSH saw nicotine as playing a role in smoking, noting that 'there are many reasons why people start to smoke but dependence on nicotine is probably the most important single reason for their continuing to smoke',[67] they investigated nicotine with other tobacco constituents as potentially harmful and therefore in need of reduction. Russell was critical of their 'low-tar, low-nicotine' approach, suggesting instead low-tar and medium-nicotine cigarettes.[68] In a comment on the ISCSH's second report, Jarvis and Russell explain that the committee ignore smokers' tendency to maintain a consistent nicotine intake by inhaling more deeply, which undermines the health advantages of switching to low-tar low-nicotine cigarettes.[69] In this network, nicotine is a contested, ambiguous actor.

ASH's approach also shifted during the 1970s. A new director was appointed whose approach was more media aware, and ASH began to work at creating news rather than just reacting to it.[70] From its creation the group had a close relationship with government, particularly with the Health Department, by whom it was initially funded, and was able to successfully apply pressure for greater control of tobacco. ASH became increasingly central in the emerging public health coalition around tobacco, and its attitude towards smoking became progressively more 'hard-line'. During the late 1970s ASH developed a close relationship with the Health Education Council (HEC) and, feeling that the focus on tobacco substitutes had held up progress, the position of both organizations on the ISCSH's 'safer smoking' programme became increasingly critical. With support from influential public health figures, in the late 1970s ASH started translating the smoking problem into a question of how to eliminate smoking. Moreover, as it became more focused on abstinence, ASH's relationship with the tobacco industry became increasingly hostile.

---

65 ISCSH, *Third Report of the ISCSH* (1983).
66 ISCSH, *Fourth Report of the ISCSH* (1988).
67 ISCSH, op. cit., n. 65, p. 5.
68 Russell, op. cit., n. 36, pp. 1430–3.
69 M.J. Jarvis and M.A.H Russell, 'Comment on the Hunter Committee's second report' (1980) 289 *Brit. Medical J.* 994–5.
70 See Berridge, op. cit., n. 2 for more detail on this period.

The stance of ASH and the HEC shaped the approach of the emerging public health network focused on tobacco. During this time the public health approach to smoking, influenced by social psychology, concentrated on changing attitudes and behaviours, and emphasized self-control in stopping smoking.[71] Consequently, the efforts of this emergent network, as it detached itself from the network around safer smoking, were directed towards health education media campaigns, control of tobacco advertising, higher taxation, and a growing focus on the rights of the non-smoker. Berridge points out that this new direction was part of a wider programme:

> The 'new public health' concentrated on relationships and the responsibility of the individual. Self-discipline, central publicity, and habit-changing campaigns were central to its ethos.[72]

Much of the push for control of tobacco was centred on advertising, the main focus of a succession of voluntary agreements; control of the product itself was managed through committee-led collaboration with industry. A regulatory regime for the tobacco products in circulation, dominated by a collaborative approach and soft-law mechanisms, was gradually put together and solidified as the health impacts of tobacco use became more widely accepted and the tobacco control coalition became more influential. Tobacco products were gradually reshaped as socio-legal objects. The influence of ASH and the increasing centrality of abstinence in policy circles meant that the concept of smoking as nicotine addiction met with resistance. A colleague of Russell's commented:

> As the tobacco control field grew, most people in the field, apart from us lot, the scientists at the unit, were very, very uncomfortable with the idea of nicotine treatment because it medicalises things ... They didn't really like the idea of emphasising that it's an addiction. In their minds it went against the idea that you can change it ...[73]

In the public health network nicotine was less a contested actor than a marginal one. This sheds light on the difficulties Russell and Fernö encountered in their attempts to extend the nicotine addiction assemblage to the wider public health network forming around tobacco. Moreover, it underlines how policy strategies shape the material, feeding in turn into regulatory frameworks and back into the 'thing'. The next section emphasizes these points by tracing nicotine gum's circulation to the British market.

71 id.
72 id., p. 179.
73 Interview three, psychologist.

With nicotine gum still an ambiguous product in Sweden, Fernö and colleagues turned their attention to other countries and submitted registration applications in Switzerland, the United Kingdom, and Canada.[74] In the United Kingdom, 'Nicorette' was launched by manufacturers Lundbeck on 16 June 1980. Prior to the launch, the CSM considered whether nicotine gum ought to be given regulatory approval. A colleague of Russell's suggested that nicotine gum was seen as 'a medicine right from the start' in the United Kingdom.[75] Its positioning within a pharmaceutical company was important:

> I think it was seen as a drug from the beginning by the inventor of the product ... he was head of the R&D of the pharmaceutical company so it was a natural thing for him to see.[76]

This suggests that, in this instance, the institutional background that became embedded into nicotine gum is what allowed it to move forward into regulatory networks. The team at the ARU had also produced further evidence on the safety and efficacy of the gum. Nevertheless, there was initial concern about the safety of a nicotine-containing medicine, particularly its cardiovascular effects: nicotine's translation into a safe substance remained unstable. Ultimately the CSM found nicotine gum to be a medicinal product that was satisfactory in quality, safety, and efficacy, and licensed Nicorette as a prescription-only medicine to be used as a 'tobacco substitute in smoking cessation'.[77]

This was not, however, the end of the story; another actor intervened in the Nicorette actor-network. Lundbeck submitted Nicorette to the Advisory Committee for Borderline Substances (ACBS)[78] to consider whether the product could be prescribed at National Health Service (NHS) expense. In correspondence between representatives of the ACBS and Lundbeck it is clear that Nicorette was being enacted differently. Lundbeck referred to it as 'our new pharmaceutical product' and stressed the activeness of nicotine.[79] The ACBS challenged this definition, raising concerns about a particular phrase:

---

74 Interview four, pharmacist at Leo.
75 Interview three, psychologist.
76 Interview four, pharmacist at Leo.
77 R.J. Anderson, 'Letter to W.P.J. Evans, 20 August 1979' (National Archives: MH149/2021).
78 A committee set up because of concerns that doctors ought to be discouraged from prescribing preparations of a 'doubtful or unethical' nature and should justify prescribing products it could be argued were not drugs. With the introduction of licensing controls over medicines in 1971, the committee's terms of reference were limited to the consideration of 'borderline substances' and it was renamed the 'Advisory Committee on Borderline Substances' (ACBS).
79 W.P.J. Evans, 'Letter to Mr Bennett, ACBS, 18 September 1979' (National Archives: MH149/2021).

'... until the ACBS has completed consideration of the drug'.
   We feel that the use of the word 'drug' in this context to be unhelpful, indeed, it may unintentionally mislead doctors.[80]

Lundbeck's Nicorette was a pharmacologically active drug and a medicine to treat a medical condition, whilst the ACBS were dealing with an uncertain object, which it was their job to define.

The ACBS formally considered Nicorette at a meeting in March 1980. In the meeting, Nicorette is labelled as an 'anti-smoking preparation'.[81] The ACBS considered four 'anti-smoking preparations' during 1973–74. A summary of their position reports that there was no evidence from controlled trials, and the products had 'no therapeutic effect on the patient's condition'; therefore they were considered 'not a drug'. The minutes also record that they were 'mindful of the importance of having the views of specialists in respiratory medicine and of epidemiologists in coming to a decision', thereby delineating other actors who would be enrolled.

Once 'expert' opinions had been collected, the committee made their decision on Nicorette in a meeting in October 1980. The minutes record the chairman's suggestion that they 'consider whether smoking can properly be seen as a disease' and the committee's conclusion that, 'smoking should be regarded more as a habit than an addiction'. It is noted that the trials the experts reviewed were found by them to be 'defective in their methodology' and the feeling of the committee was that:

> Nicorette was a nicotine substitute which did not appear to have a truly curative effect on those who used it in the hope that it would eradicate their smoking; that there was demonstrably a need for more testing of the product over a longer term with more people.[82]

For these reasons the ACBS placed Nicorette in the category: 'not a drug'.

This decision placed Nicorette in the strange position of being a licensed medicine available on prescription that doctors could only prescribe privately (meaning the patient must pay the whole cost rather than the publicly funded healthcare system). Medicines regulations enrolled Nicorette as a safe, effective medicine to treat a specific disease; however the ACBS intervened, designating smoking as a habit and translating Nicorette into an ineffective 'anti-smoking preparation' thus impeding its circulation within the network by limiting the ability of doctors to prescribe it and patients to access it. The ability of Nicorette, as a socio-legal actor, to move and act became defined by the regulatory frameworks in which it was enrolled and which came to constitute Nicorette, with their attached constraints and possibilities.

---

80 D.R. Chamberlain, 'Letter to W.P.J. Evans, 21 July 1980' (National Archives: MH149/2021).
81 ACBS, 'Minutes of meeting, 26/03/1980' (National Archives: MH149/2018).
82 ACBS, 'Minutes of meeting, 22/10/1980' (National Archives: MH149/2021).

## CONCLUSIONS

By following the translations that nicotine underwent during the 1970s we have brought into (partial) view a variety of different networks and explored some of the ways that they intersect and by-pass one another. Moreover, we have investigated the interactions between cigarettes, nicotine gum, and the regulations they encounter. It is clear that, for Russell and Fernö, nicotine was central. They translated smoking into a problem of addiction to nicotine, a disease, and it followed that, in their network, the value of nicotine gum, the treatment, was clear. Despite their efforts, a great deal more work was required to stabilize and extend this assemblage; nicotine gum remained an ambiguous product. For Leo Pharmaceuticals, nicotine gum was a chewing gum containing a poison, intended for an indication that did not exist: smoking was widely understood as a habit requiring willpower to break. Developed within a pharmaceutical company already enmeshed in networks of medicines regulation, the institutional background embedded into nicotine gum compelled it to move into regulatory networks. In Sweden the gum was initially able neither to fit the category of food nor drug and more work was required to configure nicotine gum as a medicine.

In the United Kingdom, as nicotine gum moved to new sites, it was largely ignored. The ISCSH, whose relationship with the tobacco industry likely shaped the way that they approached the problem, were focused on transforming cigarettes to make them less risky, with nicotine of only peripheral interest. The ISCSH's nicotine was a potentially harmful tobacco constituent. The growing influence of ASH and their translation of the smoking agenda so that stopping smoking was the only acceptable goal made nicotine a marginal actor in public health circles. Nevertheless, the gum was able to extend the nicotine addiction network and was translated into Nicorette: a licensed, 'prescribable drug to help people give up smoking'. The ACBS, however, intervened in this translation, designating Nicorette as a 'drug-in-the-making'. Their consultation with experts in the smoking and health field highlighted that both the definition of what sort of problem smoking was, and what sort of thing Nicorette was, remained fluid. Their decision constructed Nicorette as 'not a drug' and therefore not available on prescription within the NHS. Nicorette was reshaped as it entered new networks and could not extend the nicotine addiction/treatment network fully. Nicotine was an addictive drug, and Nicorette an effective treatment, only in certain places.

During the 1970s and 1980s the government began to use soft-law mechanisms to reorder the complex web of connections of production, distribution, consumption, and meaning that constituted the large and stable cigarette network. Public education, product modification, and collaboration with other stakeholders were emphasized. A regulatory regime for the tobacco products in circulation was gradually pieced together and solidified, and tobacco products were slowly formed as socio-legal objects. Nicotine gum, on the other hand, had to be shaped into existing categories to be enrolled into an established regulatory regime. The success of nicotine gum depended on it acting as a medicine, and since there was a pre-existing regime within which medicines were regulated, it was enrolled into this regime.

Tracing the movement of cigarettes, nicotine, and Nicorette shows how connections made in this moment of the past shape the present. Different nicotine-containing products became subject to different rules precisely because they came to occupy different categories – tobacco/medicine – and were seen as different types of thing. The circulation of tobacco and nicotine gum through separate networks, where socio-legal objects are enabled to act in different ways, increasingly stabilized cigarettes as risky consumer products to be abstained from and MN as smoking cessation medications to be carefully regulated, leaving the 'cure' more strictly controlled than the 'problem'. The category of 'medical treatment for smoking' was gradually assembled and strengthened throughout the late 1970s and 1980s, as was the idea of nicotine as the key actor linking different types of products. Therefore, only recently has 'nicotine product' become a meaningful category. Nicotine's ambiguity and multiplicity made it difficult to stabilize. Once translated by the ACBS, nicotine gum and later MN products remained as uncertain, not-quite medicines for two decades. A great deal of work was required by members of the tobacco control community to make MN prescribable on the NHS, and thus more widely available, and for the category of nicotine-containing product to be constructed. This example underlines Bowker and Star's point that classifications have real, material consequences.[83]

The use of ANT in this case study has shifted attention from stable sets of networks and highlighted the work needed gradually to put together, extend, and dismantle actor-networks. Viewing regulations as *socio-legal objects* leads one to focus on regulations as actors in heterogeneous networks that both shape and are shaped by the networks of which they are part. Regulations can be seen as playing a role in stabilizing networks and actors – enrolment into a regulatory regime stabilized nicotine gum as a medicine. However, regulation alone does not stabilize actors: in the case of nicotine gum, other actors needed to be enrolled before it could act as a medicine, whilst a broad range of actors have had to be enrolled to stabilize the tobacco regulatory regime. Nevertheless, regulation becomes a component of entities, and makes them into something specific. Materials, therefore, can also become socio-legal objects, constituted by the regulatory networks in which they are enrolled, which travel with them as scripts both enabling and constraining their actions. Cigarettes and MN are inherently 'constructed' as what they are partly by the embedding of certain regulatory frameworks and

---

83 Bowker and Star, op. cit., n. 46.

pressures into the thing itself, meaning they travel differently, are perceived differently, and influence individuals differently. It is clear that very different processes are involved when an actor confronts an established regulatory regime than when it is gradually enmeshed in an emergent one, where the reshaping of a strong network is required. The different regulatory approaches to cigarettes and MN mean nicotine becomes perceived as something quite different: a cause of addiction and disease, or as something that can and should be resisted by individuals.

ANT's insistence on unravelling the multitude of associations that constitute the world of one's actors makes giving a full ANT account, especially of historical change, challenging. In telling the story, some complex heterogeneous assemblages become punctualized and fragility and flux come to appear fixed. Nevertheless, an ANT-informed approach to regulation allows a more complex view of the development of regulation with a focus on process and context. Furthermore, a better understanding of how the governance of various nicotine-containing products evolved not only enables us to reflect on the current regulatory situation, but can help inform a discussion about how these products might be better regulated to help meet public health goals and the needs of smokers.

# The Donor-conceived Child's 'Right to Personal Identity': The Public Debate on Donor Anonymity in the United Kingdom

ILKE TURKMENDAG*

*On 1 April 2005, with the implementation of the Human Fertilisation and Embryology Authority (Disclosure of Donor Information) Regulations 2004, United Kingdom law was changed to allow children born through gamete donation to access details identifying the donor. Drawing on trends in adoption law, the decision to abolish donor anonymity was strongly influenced by a discourse that asserted the 'child's right to personal identity'. Through examination of the donor anonymity debate in the public realm, while adopting a social constructionist approach, this article discusses how donor anonymity has been defined as a social problem that requires a regulative response. It focuses on the child's 'right to personal identity' claims, and discusses the genetic essentialism behind these claims. By basing its assumptions on an adoption analogy, United Kingdom law ascribes a social meaning to the genetic relatedness between gamete donors and the offspring.*

## INTRODUCTION

In the United Kingdom, a large number of infertile couples seek *in vitro* fertilization using donor gametes.[1] According to the Human Fertilisation and Embryology Authority (HFEA) Register in 2006, 1124 children were born

---

\* Policy Ethics and Life Sciences Research Centre (PEALS), University of Newcastle, Claremont Bridge, Newcastle upon Tyne NE1 7RU, England
ilke.turkmendag@ncl.ac.uk

1 The most common assisted reproduction method. It involves removal of sperm and eggs (gametes) from the couple to produce embryos in the laboratory. If the patient cannot conceive their own gametes, donated gametes or embryos can be used. 'Donor conception' refers to the use of gametes or embryos that have been donated by a third person to enable intended recipients to become parents.

using donated eggs and sperm.[2] Using donor gametes to conceive may be as close an approximation to genetic parenthood as possible; one might be able to give birth to 'a baby', and that baby may even be genetically connected to oneself or one's partner.[3] Nevertheless it does not constitute an equivalent alternative for those for whom genetic relatedness to their children is of great importance.[4] In societies where genetic relatedness is perceived to be both a natural desire and the social norm, donor conception might invoke abnormality, and having a donor-conceived child may constitute a permanent charge of deviance against the family. Hence, most donor-conceived families have an information control strategy that will give the resultant child and parents the greatest comfort and ease so as to make the experience as 'natural' as possible. A common strategy is keeping donor conception a secret, so that the family and the child pass as 'normal'.[5] However, since the 1980s, this parenting strategy has been challenged by the disclosure regulations. Sweden was the first country to remove anonymity from gamete donors, in 1985. Since then, similar laws were introduced in a number of European countries. The United Kingdom joined this group in 2004.

Towards the end of the 1990s, there was an alarming decrease in the numbers of people coming forward as sperm donors in the United Kingdom. For example, 437 sperm donors were recruited in 1994–1995, but only 271 in 1998–1999.[6] Because of the shortage, it was claimed that some would-be parents became so desperate that they placed advertisements in newspapers. The problem was perceived to be so severe that there were efforts to establish a new independent organization to promote egg and sperm donation to infertile couples.[7] Interestingly, while donor shortage was becoming a growing concern, in 2001, the government launched a consultation to review the legislation governing access to information for those conceived through gamete donation. Given that lifting anonymity has generally had a negative impact on both the demand for, and the recruitment of, gamete donors,[8] it is

---

2 Human Fertilisation and Embryology Authority (HFEA), 'For parents of donor-conceived children' (2008), at <http://www.hfea.gov.uk/en/1185.html>.
3 M. Strathern, *After Nature: English Kinship in the Late Twentieth Century* (1992).
4 D. Elsner, 'Just Another Reproductive Technology? The Ethics of Human Reproductive Cloning as an Experimental Medical Procedure' (2006) 32 *J. of Medical Ethics* 596–600.
5 I. Turkmendag, 'The Removal of Donor Anonymity in the UK: The Silencing of Claims by Would-be Parents' (2009) PhD thesis, University of Nottingham.
6 HFEA Register (2008); 'Number of sperm and egg donors' at <http://www.hfea.gov.uk/en/1462.html>.
7 'Push for sperm and egg donors' *BBC News*, 15 June 1998, at <http://news.bbc.co.uk/1/low/health/113119.stm>.
8 J.N. Robinson, et al., 'Attitudes of donors and recipients to gamete donation' (1991) 6 *Human Reproduction* 307–9; L.R. Schover et al., 'The personality and motivation of semen donors: a comparison with oocyte donors' (1992) 7 *Human Reproduction* 575–9; R. Cook and S. Golombok, 'Ethics and society: A survey of semen donation: phase II – the view of the donors' (1995) 10 *Human Reproduction* 951–9; S. Paul et al.,

curious why the government considered a policy change which would worsen the ongoing donor shortage.

In an attempt to understand the background to the consultation process, I found social constructionist theories of social problems useful. In the following section, I introduce a few concepts borrowed from the theories of social problems. Then, I turn to the donor anonymity debate in the United Kingdom and examine the debate in the public realm through media presentations. Using a constructionist approach, I deal with the ways in which claims were formed and presented during the debate by the stakeholders, identify a number of important claims that were made by the proponents of the child's right to know, and explain how donor-conceived children came to monopolize rights language. I also present the principal claimants, the specific claims lodged against the anonymous-donation system, and the reasons why the counter-claim makers were overwhelmed. I then focus on the child's 'right to personal identity' claims that draw a parallel between adoption and donor conception, and discuss the genetic essentialism behind the adoption analogy that led to the 2004 Regulations.

## SOCIAL CONSTRUCTIONIST APPROACHES TO SOCIAL PROBLEMS

In the study of social problems, a theoretically integrated and empirically viable research tradition did not develop until the emergence of 'social constructionist' theory.[9] By rejecting the dominant conventional social problems definitions that are suggested by functionalists and value conflict theorists, Spector and Kitsuse suggested a radical change in social problems theory. Spector and Kitsuse define social problems as 'the activities of individuals or groups making assertions of grievances and claims with respect to some putative conditions';[10] thus the analyst should focus on monitoring the activities of the people who are trying to alter these putatively 'undesirable' conditions. Social problems therefore should be understood as demanding and responding activities rather than essential features. A claim is a demand that one party makes upon another (for example, demanding services, lodging complaints, supporting and opposing some governmental practice or policy) to promote a condition as a social problem. Claim-makers are the people who make claims, and audiences are the people who judge and evaluate the importance of these claims.[11]

---

'Recruitment of sperm donors: the Newcastle-upon-Tyne experience 1994–2003' (2006) 21 *Human Reproduction* 150–8.
9 H. Blumer, 'Social Problems as Collective Behavior' (1971) 18 *Social Problems* 298–306; M. Spector and J.I. Kitsuse, *Constructing Social Problems* (1977); J.W. Schneider, 'Social Problems Theory: The Constructionist View' (1985) 11 *Annual Rev. of Sociology* 209–29.
10 Spector and Kituse, id.
11 D.R. Loseke and J. Best, *Social Problems: Constructionist Readings* (2003).

Social problems are not static conditions or instantaneous events but a sequence of activities that may move through different stages.[12] Spector and Kitsuse suggest following these stages by using a 'natural history model'. They divide the natural history of a problem into several periods, each characterized by its own distinctive kind of activities, participants, and dilemmas.[13] For example, the activities of claim-makers in defining a problem are different from the activities they perform once the problem is recognized.

In this article, I use the natural history model as a frame to illustrate the way in which certain interests, assumptions, and claims led to a change in the law. While applying the natural history model, I am more interested in the discourse of claims, and less interested in presenting the sequence of activities in a chronological order. There are two reasons for this. First, the construction and legitimation of the problem is too complicated to be explained by a simplistic linear model. This is a critique of all natural history models.[14] Secondly, my research interest required understanding the nature of claims rather than providing a historical account.

While presenting the natural history of the donor anonymity problem and the claim-making activities performed, I draw on newspaper reports, web-based material, consultation documents, responses from interested organizations to these consultations, and speeches from parliamentary debates. Conrad suggests that these presentations can be called the 'public eye'. He argues that we can see the public eye as containing the lenses through which people come to understand particular problems.[15] It is claim-makers who select the lenses through which the public will come to understand the problem. In the following section, what is revealed is the way that claim-makers in the donor anonymity debate in the United Kingdom shaped these lenses by using distinct kinds of claims about abolishing anonymity.

## THE NATURAL HISTORY OF DONOR ANONYMITY AS A SOCIAL PROBLEM

Despite the fact that social problems are not characterized by a profound sense of historicity, identifying by whom and when the problem was first articulated in the public sphere may give clues about their emergence. In the United Kingdom, donor anonymity emerged as a social problem at the end of 1990s. The Children's Society (a national charity involved in campaigning and social policy work to support children) was the first organization that

---

12 id.
13 Spector and Kituse, op. cit., n. 9.
14 id.
15 P. Conrad, 'Public Eyes and Private Genes: Historical Frames, News Constructions, and Social Problems' (1997) 44 *Social Problems* 139–54.

attempted to transform the anonymity of donors into a public concern. As we shall see, later on, *Rose and Another* v. *Secretary of State for Health,* a case backed by the civil rights group Liberty, increased the pressure on the government to lift the anonymity condition.

1. *Stage 1: the Children's Society's call for a change in law*

In stage 1 of the natural history model, groups attempt to assert the existence of some condition, define it as offensive or undesirable, publicize their assertions, and turn the issue into a public or political matter. The complaining group may or may not be the victim of the said condition: for example, the complaint may be made by an organization of social workers or another humanitarian group.

In November 1998, the Children's Society called for a change in the law so that people who were born by sperm or egg donation could access the same information about their donors that adopted children could access about their natural parents. Project Manager Julia Feast said:

> There are a generation of children growing up today *who do not know who they are*. We have learned from people who have been *adopted* how important it is to *have access to medical information* so they can make informed decisions about themselves. These children's rights have been overlooked and we are sitting on a timebomb. [emphasis added][16]

The Society's characterization of the problem was significant and influential as they drew strategically on trends in adoption law. There are a number of claims in this statement: that donor-conceived children do not know who they are; that access to medical information is important to make informed decisions about one's self; that children's rights have been overlooked; and that it is only a matter of time before donor-conceived children protest against the status quo. According to the statement, donor anonymity creates identity problems for donor-conceived children who have similar needs to adopted children, and denying access to their identity infringes their rights. As we shall see later, the donor anonymity debate was characterized by this definition in the public sphere.

The Children's Society's call for legal change brought a response from government agencies. In 1999 the Department of Health confirmed that it was looking at the issue and would publish a consultation paper, although, as we shall see, it took two more years to start a consultation, but the Children's Society had successfully initiated a debate.

During the donor anonymity debate, it was not only a call by an interest organization or expert opinion that made the government act, but also a successful application for judicial review by two donor-conceived indivi-

---

16 Quoted in S. Simpson, 'Call to trace sperm donor parents' *BBC News*, 18 November 1998, at <http://news.bbc.co.uk/1/hi/health/217012.stm>.

duals. On 26 July 2002, Liberty announced the case in a press release headed 'Donor insemination case – children can claim right to personal identity'.[17] The ways in which Liberty formulated the donor anonymity problem was as important as the case itself. On its website, Liberty noted that the case sought to defend the rights of individuals to information necessary for an understanding of their 'personal identity', a similar formulation to that of the Children's Society, as we shall see later on, suggesting that genetic information is the most basic information.[18]

2. *Right to personal identity:* Rose and Another[19]

In 2002, Liberty brought Joanne Rose and EM's case to the High Court. Rose, an adult woman, had been conceived in the United Kingdom using donor insemination prior to the HFE Act. She had not been able to discover any information about the sperm donor. The other claimant, EM, a six-year-old, had been conceived using donor insemination after the Act came into force.[20] These claimants had sought access to information about their anonymous sperm donors and the establishment of a contact register. In support of their application, the claimants relied on Articles 8 and 14 of the European Convention on Human Rights (ECHR). Article 8 provides for a right to respect for private and family life, and the European Court of Human Rights has held that this right incorporates the concept of personal identity, including the right to obtain information about a biological parent.[21] The claimants also invoked Article 14 in conjunction with Article 8, arguing that there should not be discrimination between donor offspring and adoptees or between donor offspring (like Rose) born before the coming into force of the 1990 Act and those (like EM) born thereafter. The judge, Scott Baker J, outlined a series of principles based on which he concluded that 'Article 8 is engaged both with regard to identifying and non-identifying information'.[22]

The judge said that he found it:

> entirely understandable that A.I.D. children should wish to know about their origins and in particular to learn what they can about their biological father or, in the case of egg donation, their biological mother.

It was, in his view, quite clear that Article 8 ECHR and the existing jurisprudence of the European Court of Human Rights supported the idea that 'everyone should be able to establish details of his identity as a human being', and that this clearly included the 'right to obtain information about a

---

17 Liberty, press release, 26 July 2002.
18 id.
19 *Rose and Another* v. *Secretary of State for Health and HFEA* [2002] EWHC 1593 (Admin).
20 In the case, EM's mother acted as her litigation friend.
21 See, for example, *Mikuliæ* v. *Croatia* (2002) application no. 53176/99.
22 *Rose,* op. cit., n. 19, para. 46.

biological parent who will inevitably have contributed to the identity of his child'.[23] Scott Baker J's judgment says nothing however about whether there had been a *breach* of Article 8 in this case: it focuses only on the fact that Article 8 is engaged.[24] Ultimately, although Rose and EM's application for judicial review was successful at the first stage, the later hearing to determine whether there had in fact been a breach of Article 8 was delayed and, as we know, regulations passed in 2004 abolished donor anonymity.

A feature of Stage 1 in the promotion of social problems is that not only are claims formed and presented but strategies are also developed to press these claims and gain support through the creation of public controversy. Four major claims were formed in Stage 1: that donor-conceived children should have the same rights as adopted children; that donor-conceived adults have personal identity problems because of the missing information about their origins; that donor-conceived children should have access to medical information; and that donor-conceived children have a right to know about their origins. These claims created a controversy, and the government started a consultation process.

3. *Stage 2: recognition of the 'right to personal identity' as a social problem*

In Stage 2, the legitimacy of the claims is recognized by some official organizations. This may lead to proposals for reform, an official investigation or the establishment of an agency to respond.

Spector and Kitsuse note that when governmental agencies respond to the complaints of a particular group, the social problems activity undergoes a transformation. Recognition of its claim can mean that a group gets involved in official proceedings. In the donor anonymity problem, it was the Children's Society that took part in starting official proceedings. In March 2002, a MORI poll was commissioned by the Society to explore public opinion on whether children born using donated sperm or eggs should have a right to know their genetic history at the age of eighteen. MORI interviewed a sample of 1033 adults aged 15 and over throughout Britain.[25] Around two-thirds of the respondents agreed that there was too much secrecy about donor-assisted conception, and that parents should be encouraged to be more open about it; 69 per cent of the respondents were in favour of children from donor-assisted conception having the same rights as adopted children in knowing who their biological parents were, and 62 per cent agreed that donor-assisted conception should only be offered if offspring were given the right to this information at the age of 18. Arguably, this is a problematic

---

23 id., paras. 47–48.
24 id., para. 61.
25 Ipsos/MORI, 'British Public Backs Donor-Conceived Children's Rights to their Identity', at <http://www.ipsos-mori.com/researchpublications/researcharchive/986/British-Public-Backs-DonorConceived-Childrens-Rights-to-their-Identity.aspx#n1>.

question, particularly for existing donor-conception offspring, as it suggests that it might be preferable not to be born rather than not knowing one's origins. The results were welcomed by the Society:

> [t]he results from this poll are too powerful for the government to ignore. Children have been living under the shadow of legislation that has denied them the right to *the most basic information about themselves* for too long.[26] [emphasis added]

The Children's Society's efforts were supported by Baroness Warnock, one of the architects of the 1990 legislation. Warnock noted that such children should have access to information, including genetic details which could be crucial to their health. She argued that it was 'morally wrong to deceive children and deprive them of knowledge about who they are, especially when now, we all understand so much more about the importance of genetic inheritance'.[27] Her view received wide coverage in the media. Media coverage also showed that the medical community was worried about losing potential donors. It was argued that 'a majority of donors would not be prepared to donate if they thought there was a real possibility they would be identified to their donor offspring'.[28]

In February 2002, following the announcement that the legislation governing access to information for those conceived through gamete donation would be reviewed, the Department of Health published a consultation paper[29] asking what information should be available to people born as a result of gamete or embryo donation. There was widespread agreement (211 responses) that more 'non-identifying' information about donors should be made available.[30]

The government consultation ended in January 2003. A significant majority of respondents endorsed the provision of non-identifying donor information to donor-conceived children, while a smaller portion of respondents proposed the complete removal of donor anonymity.[31] Subsequent surveys that explored the views of clinics and donors suggested that ongoing donor shortage would be worsened by such policy change.[32] Experience

26 'Poll on donors favours information' *BioNews* 164, 1 July 2002, at <http://www.bionews.org.uk/page_11399.asp>.
27 'Call to end sperm donor anonymity' *BBC News*, 26 June 2002, at <http://news.bbc.co.uk/1/hi/health/2065329.stm>.
28 C. Dyer, 'Pressure increases on UK government to remove anonymity from sperm donors' (2002) 324 *Brit. Medical J.* 1237.
29 Department of Health (DoH), 'Donor information consultation: providing information about gamete or embryo donors' (2001).
30 DoH, 'Summary of Responses Received to the Donor Information Consultation' (2003).
31 E. Blyth and L. Frith, 'The UK's gamete donor "crisis" – a critical analysis' (2008) 28 *Critical Social Policy* 74–95.
32 I. Turkmendag, R. Dingwall, and T. Murphy, 'The Removal of Donor Anonymity in the UK: The Silencing of Claims by Would-be Parents' (2008) 22 *International J. of Law, Policy and the Family* 283–310.

from countries that introduced similar laws supported this. Nevertheless, the government accepted 'a strong argument in principle for children ... being able to find out the identity of their donor'.[33] It was subsequently announced, in January 2004, that people who donated eggs, sperm or embryos in the United Kingdom were to lose their right to anonymity from 1 April 2005. Anyone born using sperm, eggs or embryos donated after that date may ask the HFEA for identifying information about their donors, when they reach the age of 18. The then Parliamentary Under-Secretary of State for Health, Melanie Johnson noted:

> I firmly believe donor-conceived people have a *right to information about their genetic origins* ... including the identity of their donor (...) We live in an age where, as technology continues to develop, *our genetic background will become increasingly important*. I have listened to the views of donor-conceived people and they would like more information about their genetic origins – perhaps for themselves, perhaps for their children, perhaps because they feel the information belongs to them. That it is rightly theirs. (...) There is a growing body of opinion, which I agree with, that *donor-conceived people should not be treated so differently from adopted people*. Today's new regulations will align their positions, removing the major discrepancy that exists between the rights of donor-conceived people and those of adopted people.[34] [emphasis added]

A number of organizations, including the HFEA, the National Gamete Donation Trust, UK Donorlink, the charity Life, and Bloodlines (a pressure group campaigning for the rights of children created by sperm donation) supported the decision.

4. *Debates in parliament: a half-hearted message to the parents*

On 18 May 2004, Melanie Johnson opened the debate on the draft HFEA (Disclosure of Donor Information) Regulations 2004. She pointed out that the draft regulations were strongly supported by the Donor Conception Network, the British Association for Adoption and Fostering (BAAF), other children's organizations, and the HFEA itself.[35] Following the Commons debate, on 9 June 2004, the House of Lords discussed the issue. In her opening speech, Baroness Andrews explained the reasons that made this provision seem so necessary.[36] Some of the issues that she raised were: that the secrecy and even stigma surrounding assisted conception had faded; that public attitudes towards information and rights to information have changed

---

33 C. Dyer, 'Egg or sperm donation children will be entitled to more information' (2003) 326 *Brit. Medical J.* 240.
34 M. Johnson MP, 416 *H.C. Debs.* col. 60WS (21 January 2004).
35 Delegated Legislation Committee Debs., *Draft Human Fertilisation and Embryology Authority (Disclosure of Donor Information) Regulations 2004* (First Standing Committee) 18 May 2004, col. 6.
36 Baroness Andrews, 662 *H.L. Debs.* cols. 344–8 (9 June 2004).

dramatically (referring to the *Rose* case); and that such openness had worked successfully in recent years in relation to adoption. She argued:

> Information now is much more readily accessible than it was in 1991 ... In a century where access to information is regarded as a personal and political right, this does not seem any longer to be appropriate ... the Government are very likely to be challenged about the provision of information to donor-conceived people, as the Department of Health has already been in an application brought by Liberty.[37]

Baroness Andrews' speech indicates that by introducing the regulations, the government aimed to avoid future right-to-information cases being brought by donor-conceived children, based on Articles 8 and 14 of the European Convention on Human Rights.

However, neither the claim-makers who proposed the removal of donor anonymity, nor their opponents were satisfied by the policy change. Lifting anonymity was perceived as a positive step towards openness by the children's organizations, but it was not their ideal solution. Moreover, the shortage of gametes drove would-be parents abroad, making them vulnerable to medical risks and harm. Spector and Kitsuse[38] argue that an official response or established procedures may, in fact, turn out to be a public relations solution in which the imputed conditions are ignored on the view that the social problems activities can be 'cooled out'. Besides, even ineffective limits can send a message of disapproval and keep the state from seeming complicit in perceived immoral acts such as not being truthful with children.

## 5. *Stage 3: the re-emergence of claims*

Spector and Kitsuse[39] argue that Stage 3 is the re-emergence of claims and demands by the original group(s) or by others, expressing dissatisfaction with the established procedures. In the United Kingdom, the natural history of donor anonymity problem reached Stage 3 during a consultation exercise.[40] The child's-right-to-know camp was not convinced that lifting anonymity would encourage openness, and claims on the right to know soon re-emerged. In their response to a consultation on the HFE Act in 2005, the British Association for Adoption and Fostering (BAAF) demanded that donor conception should be registered on a child's birth records. They expressed their dissatisfaction with the new law:

---

37 id., col. 345.
38 Spector and Kitsuse, op. cit., n. 9.
39 id.
40 The Department of Health undertook a public consultation exercise over the summer and autumn of 2005 on possible changes to update the law and regulation relating to human reproductive technologies.

> We have a responsibility *not to collude with the parents who have chosen not to be truthful* with their child about his and her genetic origins. *Prospective adoptive parents would not be approved as adoptive parents if it was thought that they would not tell their child that s/he was adopted* ... this is why it should be a statutory requirement for prospective parents of a donor conceived child to undergo an assessment and preparation period ...[41] [emphasis added]

BAAF's wording in its statement is worth discussing. The advocates of an open donation system portrayed donor-conceived children not only as victims of a policy which infringed their right to identity, but also as victims of their parents, who violate moral standards by denying this information to them. Not only does BAAF construe donor conception as a form of adoption, it also suggests that there is a conflict of interest between parents and children about disclosure, and the law should not 'collude' with the parents. We can observe a similar formulation in Baroness Warnock's speech:

> There can be no moral justification whatever in *deceiving a child* about the circumstances of his birth. It is a very awkward doctrine to enunciate, considering the number of children born by adulterous relationships. Nevertheless, it is deeply morally wrong to pretend that *a child is the son or daughter of a father or mother who is not his or her real biological parent.*[42] [emphasis added]

In both statements, concealing a child's biological origins is construed as an immoral act. Such discourse attributes moral aims to the donor conception law, and suggests that the government's reluctance to compel disclosure sends a half-hearted message to parents.

*6. Claims against the law from opponents of the open donation system*

It was not only the advocates of the child's right to know who expressed discontent about the 2004 law, but also opponents of the open donation system for whom the steady decline in the number of sperm donors was a serious concern. During the debate, one of the most influential counter-claim-making groups was the British Fertility Society (BFS). Other major professional groups (such as the Royal College of Obstetricians and Gynaecologists and the British Medical Association) were also opposed to the disclosure policy on the grounds that this would have an adverse impact on donor supply. As the government did not take their views into consideration, each organization reinforced its objection during the review of the Act by the House of Commons Science and Technology Select Committee during 2004 and 2005. The BFS claimed that the short-term effect of the change in legislation had been that the cost of donor insemination had risen enormously, and the programme had effectively been removed from the

---

41 British Association for Adoption & Fostering, 'BAAF Response to Department of Health Review of the Human Fertilisation and Embryology Act' (2005).
42 Baroness Warnock, 662 *H.L. Debs.* col. 356 (9 June 2004).

NHS as standard practice in most areas.[43] Despite having supported the law, the National Gamete Trust expressed concern about the donor shortage and noted that the United Kingdom was nowhere near solving the problem.[44] The patient organization Infertility Network UK (INUK) also drew attention to the declining number of donors, leaving patients with no choice but to go abroad for treatment.[45]

*7. Stage 4: donation policy review*

Spector and Kitsuse[46] argue that Stage 4 involves rejection by the complainant group(s) of the agency's response and the development of alternatives. In the United Kingdom, the donor anonymity problem reached this stage very recently. Since the new law came into effect, fertility travel and the purchase of sperm online increasingly featured in the media. Concerns about this gradually led the regulatory body to revise its approach and policy. On 26 August 2010, the authority announced its decision to hold a full public consultation into its policies on expenses, compensation, and benefits in kind donors can receive for donation and the number of families donors can donate to. Although the government has made it clear that it does not intend to reverse the removal of donor anonymity,[47] the recent policy consultation indicates a potential shift in United Kingdom policy.

## HOW DID CHILDREN'S RIGHTS COME TO MONOPOLIZE RIGHTS CLAIMS?

The social constructionist approach to social problems suggests that any social condition is a potential subject for claim making, but it is claim-makers who shape the public's sense of what the problem is. The natural history of the donor anonymity problem indicates that during the debate that led to the 2004 Regulations, without any significant objection from would-be parents, the donor anonymity problem came to be owned by advocates of the child's right to know. My analysis also suggests that the decision to abolish donor anonymity was strongly influenced by a discourse that asserted donor-

---

43 British Fertility Society (BFS), 'BFS Response to the Department of Health Review of the Human Fertilisation and Embryology Act' (2007).
44 M. Henderson, 'Sperm donor figures rising despite loss of anonymity' *Times*, 4 May 2007.
45 'Infertility Network UK Response to HFEA Press Release On Latest Figures For Sperm Donor Recruitment' *Medical News Today*, 3 May 2007, at <http://www.medicalnewstoday.com/articles/69785.php>.
46 Spector and Kitsuse, op. cit., n. 9.
47 DoH, *Review of the Human Fertilisation and Embryology Act: Proposals for revised legislation (including establishment of the Regulatory Authority for Tissue and Embryos* (2006; Cm. 6989).

conceived children's 'right to personal identity'. The view that donor-conceived children have a right to access identifying information was legitimized by two principal claims: that genealogical knowledge is central to the development of personal identity; and that the position of donor-conceived people should be aligned more closely with that of adopted people with access to identifying information about their 'biological parents'.[48]

During the debate the potential impact of the disclosure policy on would-be parents has been raised as a social problem, but in the United Kingdom, would-be parents neither mobilized around access-to-treatment claims, nor did they seek to construe these claims as a human right. Hence, until recently, donor shortage has not become a social problem that calls for regulatory response. The would-be parents' reluctance to mobilize around pressing claims to protest the removal of donor anonymity partly reflects the variety of ways in which they can avoid the impact of this legislation. If the parents do not agree with the law, they can keep donor conception secret. They also have readily available alternatives in treatment in other medically-advanced countries.

Nevertheless, counter-claim activities were performed by the medical community and the patient support group, as shown above.[49] However, the fact that would-be parents would suffer because of the donor shortage could find only little sympathy when weighed against the child's welfare arguments raised by the child's-right-to-know camp. As a result, would-be parents' right to seek treatment was compromised for the perceived needs of donor-conceived children. As we shall see in the next section, the assumptions about the needs of children conceived by donated gametes were based on the adoption analogy.

1. *The 'typification' of children conceived by donation*

Typifying is the backbone of claim-making activity. It occurs 'when claimsmakers characterise a problem's nature'.[50] Typification can take many forms. To give an orientation toward a problem and to argue that a problem is best understood from a particular perspective is one form. When a new problem is constructed as a different instance of an already existing problem, the audience will be more likely to understand the problem: successful claim making changes the old in novel ways.[51] The adoption analogy that Children's Society used is an excellent example.

---

48 It was also suggested that donor-conceived children should know their donor to ascertain their medical history, albeit that medical predictions can be made just as accurately from one's own body.
49 'Sperm donors to lose anonymity' *BBC News,* 21 January 2004, at <http://news.bbc.co.uk/1/hi/health/3414141.stm>.
50 J. Best, *Images of Issues: Typifying Contemporary Social Problems* (1995) 8.
51 D.R. Loseke, *Thinking about Social Problems: An Introduction to Constructionist Perspectives* (2003).

On 26 June 2002, the *Observer* newspaper published a commentary by Julia Feast of the Children's Society, perfectly exemplifying how the adoption analogy was employed:

> This Sunday, millions of families throughout Britain will celebrate Fathers' Day – a day to celebrate and honour the importance of family values. Yet the 1500 children born each year through donor-assisted conception *are denied the rights to even know who their biological parents are.*[52] [emphasis added]

Here, the importance of family values is emphasized, and donors are conceptualized as 'biological parents'. Feast portrays donor-conceived children as victims of a policy which denies them rights that adopted children enjoy. As discussed earlier, Feast also construes access to identifiable information about the donors as children's 'right to identity':

> It's time for the government to acknowledge that openness and honesty should now become the accepted practice, so that all of tomorrow's children grow up with dignity and a *right to their identity*.[53] [emphasis added]

Here Feast addresses humanitarian values such as 'openness', 'honesty', and 'dignity', and draws attention to the conflict between these humanitarian mores and government policy. She goes on to note that by denying them access to donor identity, the government infringes human rights in the 'European Convention on the Rights of a Child' [*sic*]. According to Feast, the right to information about identity is also 'central to mental health'.[54] This typification enables Feast to construe a distinctive social group: children who have weak identity because of the missing information about their birth parents. Apparently, when the public is convinced of similarities between adoption and donor conception, they will know how to think about the problem. Typification of adopted people as 'others' has also been a strategy of 'search movement' activists in America:

> The agenda of the search movement draws upon society's view of adoptees as simultaneously familiar, to the extend that the public can identify with their quest for identity, and different, to the extend that they are perceived as standing outside the order of nature, as Others.[55]

The child's-right-to-know claims seem to adopt ideas from social movements, particularly the 'search movement' for adopted people.[56] Studies of adopted children have found that a child's knowledge of his or her

---

52 J. Feast, 'The right to an identity' *Observer*, 26 June 2002.
53 id.
54 id.
55 K. Wegar, *Adoption, Identity, and Kinship: The Debate Over Sealed Birth Records* (1997).
56 Wegar refers to adopted children's rights movement to access to sealed birth records in United States (from the 1970s) as the 'search movement'.

background is crucial to positive and non-confused self-identity.[57] On this basis, it is argued that disclosure is in the best interests of the child.[58]

## 2. The adoption analogy and disclosure

Using the adoption analogy, the Children Society typified the emotional and psychological needs of the children conceived by donor conception, and construed the need for identity as a familiar quest with which the public can identify.

However, the parallel between adoption and donor anonymity is questionable. Adoption is a family-building activity that involves pre-existing individuals, whereas donor conception is directed towards creating a child in order to create a family.[59] In other words, adoption is a *substitute* for procreation whereas donor conception is a *form* of procreation: the act has its own integrity and completeness – it is the would-be parent(s)' act and the child is unquestionably their child. 'The act of procreation which takes place by artificial insemination is undoubtedly the act of the couple, and more particularly of the mother.'[60] In this act, social links are established between the procreator and the child, not with the donor.

In adoption, the relation of the child with the birth mother/father has a social context, and there is a relinquishment that calls for explanation. Unlike adoption, in donor conception the resultant child did not experience a relationship with a genetic parent that was later broken.[61] In other words, in donor conception, genetic linkage does not represent a social 'relatedness' in the same way as adoption; children conceived by donation are linked to their donors by shared genes, not by a birth story.

Besides, in donor conception, openness is not always thought to be the best for the child's interest. Some parents who have donor-conceived children seem to be more concerned with protecting the child from the potential harm of knowing the truth rather than giving the child greater autonomy at any cost.[62] Although secrecy may damage the relationships, the disclosure of reproductive secrets often has detrimental effects. Smart argues

---

57 J. Triseliotis, *In Search of Origins* (1973); A. McWhinnie, 'Gamete donation and anonymity: Should offspring from donated gametes continue to be denied knowledge of their origins and antecedents?' (2001) 16 *Human Reproduction* 807–17.
58 S. Maclean and M. Maclean, 'Keeping Secrets in Assisted Reproduction – the Tension between Donor Anonymity and the Need of the Child for Information' (1996) 8 *Child and Family Law Q.* 243–51.
59 E. Haimes, 'The making of "the DI child": changing representations of people conceived through donor insemination' in *Donor Insemination: International Social Science Perspectives*, eds. K.R. Daniels and E. Haimes (1998) 53–75.
60 id.
61 S. Golombok, 'Anonymity – or not – in the Donation of Gametes and Embryos' in *Regulating Autonomy: Sex, Reproduction and Family*, eds. S. Day Sclater, F. Ebtejah, E. Jackson, and M. Richards (2009) 223–38.
62 Turkmendag et al., op. cit., n. 32.

that 'the problem with knowledge about genetic links is that it is not mere "information" but it is powerful knowledge that changes relationships regardless of the wishes those involved'.[63] Such fear may explain why a significant majority of donor conception parents have tended not to disclose the fact to their children. A recent study shows that less than 8 per cent of egg-donation parents, and less than 5 per cent of those who used donor insemination, disclosed to their children.[64] Studies show that the predominant motive is to avoid potential difficulties in the relationship between the non-genetic parent and the child.[65]

Information that is currently available suggests that the psychological well-being of children is not compromised by secrecy.[66] This suggests that disclosure should not be imposed on families. But there is a remarkable difference between disclosing to children their means of conception, and imposing on them that their genetic relatedness to the gamete donor is an indispensable component of their personal identity, and that development of their identity may depend on finding identifiable information about their 'real biological parents'.

## DISCUSSION

Through examination of the donor anonymity debate in the public realm, and adopting a social constructionist approach, this article focuses on how donor anonymity was defined as a social problem requiring a regulative response. The natural history of the donor anonymity problem suggests that in the United Kingdom, adoption was regarded as a morally and psychologically relevant model for approaching the need to find out identifiable information about their origins claimed by some children. Case law often uses analogies. On the other hand, as Nelson argues, 'whether the analogy will seem convincing depends on the kind of normative light in which one is standing'.[67] Accordingly, if one's 'primary source of moral illumination' is that the donor is 'personally present through his genetic contribution',[68] one

---

63 C. Smart, 'Law and the Regulation of Family Secrets' (2010) 24 *International J. of Law, Policy and the Family* 397–483, at 397.
64 S. Golombok et al., 'Non-genetic and non-gestational parenthood: consequences for parent-child relationships and the psychological well-being of mothers, fathers and children at age 3' (2006) 21 *Human Reproduction* 1918–24.
65 id.; J. Burr and P. Reynolds, 'Thinking ethically about genetic inheritance: liberal rights, communitarianism and the right to privacy for parents of donor insemination children' (2008) 34 *J. of Medical Ethics* 281–4; Golombok, op. cit., n. 61; Turkmendag, op. cit., n. 5.
66 Golombok, id.
67 H.L. Nelson, 'Dethroning Choice: Analogy, Personhood, and the New Reproductive Technologies' (1995) 23 *J. of Law, Medicine & Ethics* 129, at 130.
68 K. O'Donovan, 'A right to know one's parentage?' (1988) *International J. of Law, Policy and the Family* 27–45, at 43.

will find the adoption analogy convincing. If, on the other hand, we regard the sperm and egg as a human material but not personally human, then the adoption analogy is less convincing.

The new reproductive technologies have transformed the traditional understandings of relatedness and lineage. For example, the understanding that blood relations usually reflect the genetic relationship had changed by the end of the twentieth century:

> The maternal genetic material, including the determinants of the fetus's blood type and characteristics, is contributed by the egg, which is derived from the ovaries of one woman. Nonetheless, the embryo grows in and out of the substance of another woman's body; the fetus ... takes form from the gestational woman's blood, oxygen, and placenta. It is not unreasonable to accord the gestating mother a biological claim to motherhood. Indeed, some have suggested that shared substance is a much more intimate biological connection than shared genetics and is more uniquely characteristic of motherhood, as genes are shared between many different kinds of relations.[69]

Notions of personhood can be entangled in or disentangled from human biological material.[70] Whilst some forms of human material are aligned with personhood, such as an embryo, some are categorized as residual tissue (for example, placenta, amniotic sac). However, the law often fails to recognize how representations of the human body and its parts are constructed.[71] The new United Kingdom law geneticizes the family: 'social forms like family, parenting and fatherhood are fundamentally undermined by a biological determinism in their authenticity and this is not considered within current policy'.[72] Donor conception is an approximation to genetic parenthood. It makes it possible for would-be parents to enjoy the cultural and social practices of gestation: having a 'bump', carrying the baby and feeling the movements of it, monitoring foetal development, sharing scanned images with friends and family and, finally, going into labour. But more importantly, donor conception enables infertile people to conceive a child to whom they are genetically related. Recent research shows that gestation does influence the genetic make-up of an embryo and foetus. Hence, those who conceived by gamete donation have a biological claim to parenthood. The adoption analogy of the new law disregards would-be parents' biological claim to parenthood. Clearly, however, the approximation to biological parenthood through donor conception is not straightforward: it is different from the conventional understanding of 'natural' parenthood, and raises complex questions of filiation.

---

69 C. Thompson, *Making Parents: The Ontological Choreography of Reproductive Technologies* (2005).
70 K. Hoeyer et al., 'Embryonic Entitlements: Stem Cell Patenting and the Co-production of Commodities and Personhood' (2009) 15 *Body & Society* 1–24.
71 N. Pfeffer and J. Kent, 'Framing Women, Framing Fetuses: How Britain Regulates Arrangements for the Collection and Use of Aborted Fetuses in Stem Cell Research and Therapies' (2007) 2 *BioSocieties* 429–47.
72 Burr and Reynolds, op. cit., n. 65, p. 283.

If gamete donors should be considered biological 'parents', and using donated gametes to conceive is a form of pre-adoption, then our approach to donor conception is misguided. First, we cannot assume that the donor-offspring relationship is one-sided. How can we ask someone to give away their potential baby to be adopted by strangers? How many donors are 'altruistic' enough not to worry about being contacted by their 'biological child' eighteen years later? Secondly, if using donated gametes risks the resultant child's ability to develop a sense of identity, making 'identifiable' information available is a remarkably simplistic solution. Even if the parents choose to inform the child about the donor conception, and the child obtains the identifiable information, there is no guarantee that the information available will satisfy the child. We cannot guarantee that the gamete donor will meet the child, either. Accordingly, if adoption analogy is right, donor conception raises severe ethical questions.

Is donor conception a form of adoption that should be avoided at any expense? The dilemma would seem to be easy to resolve: donating one's gametes does not make one a parent. As O'Donovan argues, bodily spare parts are transferrable and exchangeable because, 'although they belonged to someone in particular, and grew at the behest of his genetic constitution, they do not convey his genetic individuality'.[73] Donating gametes to help someone to achieve pregnancy is not congruent with the social role of parenthood.[74] Donor conception is the would-be parents' act: 'the parent must be the one who desires to be a parent'.[75] But, as Strathern points out, there is still equivocation, given that they do not supersede the 'genetic parents'.[76]

In the United Kingdom, the filiations created by donor conception have been largely overlooked in framing legislative responses to the donor-conceived child's quest for identity. One of the aims of the new law was to ease the remaining stigma around donor conception. Ironically, by marginalizing donor-conceived children, and enforcing a deeply-rooted view that genetic linkage is indeed very important, it is possible that the United Kingdom's disclosure policy compounds stigma and increases subterfuge rather than openness. The current regulation is enforcing a genetic essentialist view of filiations; it ascribes social meaning to the passage of genes, and sends society a disputable message that contributing one's genes makes one a 'parent'.

---

73 O'Donovan, op. cit., n. 68.
74 J. Wallbank, 'The role of rights and utility in instituting a child's right to know her genetic history' (2004) 13 *Social and Legal Studies* 245–64, at 260.
75 Strathern, op. cit., n. 3, p. 178.
76 id.

# A Socio-legal Analysis of an Actor-world: The Case of Carbon Trading and the Clean Development Mechanism

EMILIE CLOATRE* AND NICK WRIGHT**

*This article reviews the Kyoto Protocol's Clean Development Mechanism (CDM), and analyses how it reflects a particular international vision of climate change and its solutions. It discusses how the expectations this approach embeds have become challenged by practice, and practitioners, and how alternative models for the CDM have been put forward. The article argues that these challenges and alternatives can be understood better by borrowing Michel Callon's concept of 'actor-world', in order to analyse how contrasting visions of technologies also inevitably entail conflicting ideas about the world.*

## INTRODUCTION

Scholars working at the intersection of law, sociology, and environmental science are examining the shape and dynamics of how societies are organized so as to understand the problem of global climate change.[1] Climate science is ordered in networks ranging from global systems to local nexus.[2] In a classic

---

\* Kent Law School, University of Kent, Canterbury CT2 7NZ, England
e.cloatre@kent.ac.uk
\*\* School of Veterinary Medicine and Science, University of Nottingham, Sutton Bonington Campus, Nottingham LE12 5RD, England
nick.wright@nottingham.ac.uk

The original research for this article was funded by a grant from the Centre for Environmental Law at the University of Nottingham.

1 For an analysis and critique of the relationship between environmental law and methodological and theoretical engagement, see L. Fisher, B. Lange, E. Scotford, and C. Carlarne, 'Maturity and Methodology: Reflecting on How to Do Environmental Law Scholarship' (2009) 21 *J. of Environmental Law* 1–38.
2 For example, see A. Block, 'Topologies of Climate Change: actor-network theory, relational-scalar analytics and carbon market overflows' (2010) 28 *Environment and Planning D: Society and Space* 896; M. Callon, 'Civilizing markets: carbon trading between in vitro and in vivo experiments' (2009) 34 *Accounting, Organizations and*

Actor-Network Theory (ANT) story about the failed project of introducing new electric vehicles into 1970s France, Michel Callon put forward the interrelated concepts of 'actor-worlds' and 'engineer-sociologists'. Whilst not much drawn upon in later well-known ANT studies,[3] these two concepts are a potentially useful and interesting way to explore conflicts, resistance, and revisions generated around climate-change legislation.

This article analyses the deployment of the Clean Development Mechanism (CDM) – the most ambitious carbon-offsetting scheme to date – of the international climate agreement known as the Kyoto Protocol. The various practitioners involved in bringing the CDM into being are found to propose and deploy contrasting social visions. As such, the study is in part a response to calls for more attention to be given to the processes deployed for 'rewriting capitalism' with regards to climate-change law and the carbon market more specifically.[4]

## THE MAKING OF ACTOR-WORLDS

In his study of the electric vehicle, Callon describes how engineers at Electricité de France (EDF) proposed the development of a new type of technology, whilst offering a form of expert vision of what French society was about to become.[5] In proposing the electric vehicle, not only did they elaborate the technical aspects, but they also imagined which type of society this would fit in – which Callon refers to as an 'actor-world'. In this actor-world, EDF is expecting a certain number of actors and structures (or 'actor-networks') to fall into place, act, and interact in particular ways.

Callon's story then becomes one of failure as follows: (i) consumers do not seek the post-industrial society envisioned by EDF, and (ii) the various technical devices behave in ways other than that expected by EDF. But Callon's story is not exclusively one of failure and discrepancy but also a story about competing actor-worlds; how others of those he calls 'engineer-sociologists' propose alternative stories of viable technologies and their

---

*Society* 535; D. MacKenzie, 'Making things the same: Gases, emission rights and the politics of carbon markets' (2009) 34 *Accounting, Organizations and Society* 440; L. Lohmann, 'Toward a different debate in environmental accounting: The cases of carbon and cost-benefit' (2009) 34 *Accounting, Organizations and Society* 499.

3 Whilst engaging with the concept of actor-world, and the implied idea of co-production of socio-technical networks, this paper does not claim to be an 'ANT analysis' of the CDM.
4 For examples, see MacKenzie, op. cit., n. 2.
5 M. Callon 'The Sociology of an Actor-Network: the case of the electric vehicle' in *Mapping the Dynamics of Science and Technology*, eds. M. Callon, J. Law, and A. Rip (1986) 22; see, also, M. Callon, 'Society in the Making: The Study of Technology as a tool for Sociological Analysis' in *The Social Construction of Technological Systems*, eds. W. Bijker, T. Hughes, and T. Pinch (1989) 87.

surrounding, projected societies. In his case study, engineers at the car firm Renault do not share EDF's version of society. Where EDF's actor-world was based on the expectation that French consumers would need to abandon their current attachment to petrol-driven motor vehicles, in their quest to respond to the perceived threats of industrialization, Renault's engineers developed a vision based on the transformation of the combustion engine to make it more suitable to the demands and needs of current French society. They have different expectations, which they translate in different technologies and endeavours.

The story offered, and the core concepts deployed, bring many insights that echo the empirical findings of the present study. It helps us understand how practitioners, when developing technological models (and we understand technology here in a broad sense, to encompass, for example, legal technologies), also develop a series of assumptions about the world. In turn, it allows us to consider how these actor-worlds may fail to become realized in the form expected by their creators, or planners.

Central to this analysis is the role of those Callon calls 'prime movers'. In the case of the electric vehicle, these prime movers are engineers, or as Callon calls them, 'engineer-sociologists' – EDF engineers and Renault engineers. For him, the engineers produce competing versions of society and, in the process, create visions and understandings of society that must compete with those offered by others, including sociologists. In our case study, we explain how the actor-world of the CDM generated by lawyers, state negotiators, and UN officials, who might be termed administrator-sociologists in the context of the Kyoto protocol, has had to face alternative visions, not only of what the CDM as complex socio-legal technology could or should be, but also of the actor-world of which it is part.

In his analysis, Callon does not engage specifically with the role of the legal in shaping actor-worlds. In our article, lawyers and legal tools are a critical part of the story. We refer to them throughout and engage, in the conclusion, with the nature and meaning of the 'materialization' of law. In the next section, their understanding of the CDM is described.

## THE INTRODUCTION OF THE CLEAN DEVELOPMENT MECHANISM

In 1997, member states of the United Nations signed the Kyoto Protocol to the United Nations Framework Convention on Climate Change (UNFCCC), and agreed on a range of detailed mechanisms as a response to the problem of climate change. One of these mechanisms was the CDM, a hybrid socio-legal and technical solution that enables:

> a country with an emission-reduction or emission-limitation commitment under the Kyoto Protocol ... to implement an emission-reduction project in developing countries. Such projects can earn saleable certified emission

reduction (CER) credits, each equivalent to one tonne of $CO_2$, which can be counted towards meeting Kyoto targets.[6]

The rules elaborated within international law suggest a particular projection of how the CDM (and the carbon market) could work, in an actor-world inhabited by abstractly conceptualized, and 'simplified' elements (as pointed out by Callon in the case of the electric vehicle).[7]

The CDM is a peculiar figure of the Kyoto Protocol, in that it claims to bring together the broad and complex aims of mitigating climate change and supporting sustainable development, and to offer opportunities for technology transfer and capacity-building.[8] This has also made it one of the most controversial aspects of the protocol. In large part, this is because it works in ways which deviate from what its creators seemed to have envisioned and from the way many others would like it to work. The CDM is largely predicated on the creation of a new market in carbon off-sets and, in this way, the environment is the supposed main beneficiary in terms of less carbon emission. Projects are to be created from which emission reductions will be obtained. However, if the sustainability objective is set more widely, to include, for example, lifting people out of poverty, then the task for a 'successful' CDM project is clearly different. One of the underlying causes of dispute about the CDM is regarding the extent to which market-based mechanisms can identify and meet these needs or whether a 'very radical' political-economic vision is required where more equitable and efficient use of energy is the 'first objective of policy, rather than the supposed beneficiaries of more economic growth'.[9] Many disputes about the CDM centre around a favoured form of market tools embedded within a particular vision of the concept of a global economy.[10] It is therefore essential to explore this a little if we wish to understand the resistance exercised by particular groups, before returning to the 'everyday' difficulties associated with translating the projected CDM into practice.

The first thing to say is that disputes over political-economic development trajectories are not often recognized as such. Hulme, for example, talks about

---

6 The CDM was created in Art. 12 of the Kyoto protocol; the text of the protocol can be accessed at <http://unfccc.int/resource/docs/convkp/kpeng.pdf>.
7 Modalities and procedures for the CDM were agreed by the Conference of the Parties in Montreal in 2005 and can be found at <http://unfccc.int/resource/docs/2005/cmp1/eng/08a01.pdf#page=6>.
8 Art. 12 of the Kyoto Protocol:
   The purpose of the clean development mechanism shall be to assist Parties not included in Annex I in achieving sustainable development and in contributing to the ultimate objective of the Convention, and to assist Parties included in Annex I in achieving compliance with their quantified emission limitation and reduction commitments under Article 3.
9 M. Redclift and G. Woodgate, 'Sustainability and social construction' in *The International Handbook of Environmental Sociology*, eds. M. Redclift and G. Woodgate (1997) 55, at 58.
10 MacKenzie, op. cit., n. 2.

'the trade-off between the goal of supplying cheap emissions credits and the promotion of sustainable development',[11] rather than analysing the version of the latter implied by the former. Others are a little more explicit. Giddens is concerned that the CDM does not address issues of equity as 'it allows developed countries to relax their own emission reduction efforts',[12] whilst the fulfilment of its official aims in relation to renewable energies and technology transfers are doubtful. Backstrand and Lovbrand are explicit about the political implications of the CDM, and point out the power issues that it builds upon:

> from a critical international relations perspective the carbon offset market epitomizes continued neoliberal governance building on a capitalist compact between business and government elites in industrialised and developing countries.[13]

Finally, in Lohmann's views, carbon market mechanisms have become less fundamentally about environmental concerns and more centrally about fostering the market itself and the interests involved.[14]

These different perspectives illustrate some of the most fundamental concerns that underlie the CDM, and explain where some of the scepticism, widely spread across NGOs and other climate actors, finds its root. The CDM is a highly politicized project, which adopts a specific model of capitalism that many would reject, and have possibly started to do in practice, as we will illustrate below.

## MATERIALIZING AND DE-MATERIALIZING THE CDM

As well as these fundamental political questions, many other elements need to come together if this mechanism is to deploy as envisaged by administrator-sociologists, such as: a series of purpose-built interactions which enable 'technology transfer'; the development of norms and procedures that support these interactions; the working of these new strategies towards the fluid concept of 'sustainable development'; the progressive disappearance of some greenhouse gases through a better lining up of technologies and action. A UN body known as the Executive Board (EB) oversees the CDM. This system is very heavily organized around a certain idea of bureaucracy, and assumptions about the fact that appropriate ordering will succeed at each step of the process. A closer look at the procedures involved in the CDM and

---

11 M. Hulme, *Why We Disagree About Climate Change* (2009) 277.
12 A. Giddens, *The Politics of Climate Change* (2009) 190.
13 K. Backstrand and E. Lovbrand, 'Planting Trees to Mitigate Climate Change: Contested Discourses of Ecological Modernization, Green Governmentality and Civic Environmentalism' (2006) 6 *Global Environmental Politics* 50, at 70.
14 L. Lohmann, 'Uncertainty Markets and Carbon Markets: Variations on Polanyian Themes' (2010) 15 *New Political Economy* 225, at 247.

created within and following the Kyoto Protocol gives us a better sense of what is termed in French as *agencements* – the connections that are assumed to take place in the actor-world generated at Kyoto.

The CDM is initiated by project developers, who put forward their proposal in a Project Design Document (PDD). This newly materialized object can then travel across the various bureaucratic and technical networks of the CDM before being deployed as a project. The PDD needs to be approved by national offices within the developing country expecting to host the project (the Designated National Authority, or DNA). It will then go through the more problematic stage of acquiring validation from a commercial auditing company registered by the UN as a Designated Operational Entity (DOE). A project that has been validated is then expected to be registered by the EB. This will be subject to its offering a methodology to calculate greenhouse gas emission reduction that is approved by the UN; this is a problematic stage, in practice.

Once the project is deployed, another range of actors are expected to fall into place. These include engineers, expected to carry out the project and, in the process of doing so, to contribute to technology transfer, capacity building, and sustainable development – none of which are unproblematic actions or concepts. In addition to this, auditors and bureaucrats remain involved throughout the process, as further monitoring is expected to take place in accordance with the methods identified in the registered PDD. In the deployment of the CDM as projects, the model built by the UN is materialized as a specific set of technologies and associated practices.

The ultimate translation that a CDM project will undergo results in the production of CERs – units that can be traded on the carbon market and are both the financial outcome of the process, and the realization of the second aim of the CDM (to assist industrialized countries and countries with economies in transition in achieving their targets). At this point, the CDM is turned back from a material project to an intangible unit, in a process of de-materialization. Assuming that an audit company verifies the emission reduction, then these emissions will be certified and CERs issued. They can then be traded.

In the version of the CDM projected by the Kyoto Protocol, a number of assumptions are made, about the type of socio-technological issue that climate change is, and about the nature of the response offered by the international community in general, and the CDM in particular. First, the CDM encompasses the notion that climate change can be managed through quantification strategies – measurements, credits, valuations, are all part of a system in which the translation of a specific form of climate science into policy and practice is clear.[15] Second, it fully embeds the idea that climate

---

15 For a specific engagement with the role of auditing in social organization, see M. Power, *The Audit Society: Rituals of Verification* (1997).

change management is a global problem, and that we need to respond to it as a global society. The issues raised by this framing have been elaborated on at length elsewhere.[16]

From the design of CDM projects to their final implementation, the attribution of credits and their trading, the problem of climate change is transformed through a series of translations that turn it from a complex human/nature problem into a quantifiable matter. The CDM, and the carbon market as a whole, is about 'making things the same', in the words of MacKenzie. However, the series of translations involved are filled with uncertainty and approximations, generating an inherent fragility in the system.[17] In this chain of translation, technologies and quasi-technologies play a significant part. Knowledge must be of a nature that can be broken down and quantified, which is particularly crucial for the design of projects and their progressive deployment.[18]

The deployment of this complex system of approval and monitoring mutually supports the vision, or actor-world, that has been put forward by a certain category of practitioners – lawyers, bureaucrats, and politicians – in putting together a socio-legal answer to climate change, in the form of the Kyoto Protocol and its market mechanisms. Whilst this creates an abstract model, it cannot provide a strict and pre-defined framework that will be followed without transformations and challenges in practice. In this project, we focused on these transformations, and on the responses that administrator-sociologists deployed when faced with what they saw as the limitations of the CDM.

In our empirical study, the majority of our interviewees were identified from a 'List of Projects with UK Approval' published by the United Kingdom's Designated National Authority for the CDM.[19] In total, we

---

16 See S. Shackley, P. Young, S. Parkinson, and B. Wynne, 'Uncertainty, complexity and concepts of good science in climate change modelling: Are GCMs the best tools?' (1998) 38 *Climatic Change* 159, or C.A. Miller, 'Climate Science and the Making of a Global Political Order' in *State of Knowledge: the Co-Production of Science and Social Order*, ed. S. Jasanoff (2004) 46. In the specific context of the CDM, see E. Boyd, 'Governing the Clean Development Mechanism: Global rhetoric vs local realities in carbon sequestration projects' (2009) 41 *Environment and Planning A* 2380.
17 Mackenzie, op. cit., n. 2; L. Lohmann, 'Marketing and making carbon dumps: Commodification, calculation and counterfactuals in climate change mitigation' (2005) 14 *Science as Culture* 203.
18 On the links between the production of knowledge in the context of climate change and processes of governing and power, see A. Oels, 'Rendering Climate Change Governable: From Biopower to Advanced Liberal Government?' (2005) 7(3) *J. of Environmental Policy and Planning* 185; Backstrand and Lovbrand, op. cit., n. 13; S. Rutherford, 'Green governmentality: insights and opportunities in the study of nature's rule' (2007) 31 *Progress in Human Geography* 291.
19 DEFRA, *Designated National Authority for the CDM (UK DNA) List of Projects with UK Approval of Participation* (2009).

contacted 141 companies by email and as a result conducted nine follow-up interviews by telephone. A further five interviews were conducted with individuals identified from websites and personal contacts. We were not granted an interview with any of the major NGOs, but instead analysed publicly available documents and commentaries. Interviews were conducted with three environmental consultants, two civil servants, the chairman of a carbon finance company, an employee of an oil industry trading company, a director of a carbon trading company, an employee of the trading arm of an international bank, a lawyer at a international law firm, the vice-president of a carbon finance company, a senior policy officer of a social enterprise, a manager of an auditing company, and a development consultant. From these interviews, the many difficulties that the model projected by the UN meets in practice became clear. In addition to this, the reframing of the CDM by administrator-sociologists willing to challenge the assumptions produced by this particular model was put forward as an alternative to this original actor-world, and as a form of social critique of the CDM. In the remainder of this article, we observe some of these movements of failure, transformation, and resistance.

## PERFORMING THE ACTOR-WORLD

The empirical findings from this research suggest that the deployment of the UN actor-world met three main sets of difficulties, or challenges. First, the envisioned CDM network fails to deliver the smooth series of connections that such a system is expected to rely upon. Second, some actors find themselves excluded whilst others from within corrupt the system away from what the dominant actors envisaged. Finally, there are other actors who contest the very existence of the actor-world, its adequacy to the problems of climate change, and so other, competing actor-worlds are acted out, including by some practitioner-sociologists, who develop an alternate vision not only of the CDM but also of its surrounding networks. In this section, we will explore aspects of these challenges. Throughout, some of the critiques made of the CDM by commentators are echoed in its daily deployment.

*1. Exclusion*

In his analysis of the electric vehicle's actor-world, Callon insists that actor-worlds always project a simplified version of reality, and the complexity of enlisted elements can never be fully envisioned abstractly. In the case of the CDM, the discrepancies between the expectations built around both the 'enlisting' process and 'procedural steps', and the stories of the deployment of the CDM seem particularly stark.

One of the core examples that we found of limits and discrepancies of the actor world proposed by the UN is the fact that the enlisting of projects and the various actors that they involved appeared exclusionary. In particular, the processes of approval of the CDM mean that there are many projects and technologies which find it harder than others to be enlisted, regardless of their potential in achieving the aims of the CDM.[20] This is due to what can be understood as transformations, disconnections, and failures. Central to these are the relationships between technologies, methodologies to calculate greenhouse gas emission reductions, and the translation and transformations of these methodologies into practice. As an example, in the following quote, a project developer explained how the translation by the UN from a specific methodology to a more general one resulted in serious difficulties in getting a particular project registered. Here, the difficulties of establishing a balance between fluidity and specificity in the complex CDM networks is highlighted, in a way that cannot transpire explicitly from the original script set up in the Kyoto protocol:

> We replaced it [a coal boiler] with a gas turbine, which ... massively raises the efficiency of the plant ... and we got CERs for it ... it's a very complicated methodology ... and [it] was very specific to the source of fuel that we were using. And when the UN approved the methodology they wanted the methodology to be more widely applicable so that it could be used by projects in other countries. So without really looking at the reason for why the methodology was very specific they changed very slightly some of the terms in the methodology which meant that when we came to register our, validate our project using the methodology the validators said well hang on this doesn't all stack up and so we're now in this process where we had to go back to the UN and say look we've changed our methodology it doesn't work, what you've done isn't applicable to this project anymore and this is the methodology designed for this project.

The UN, probably for reasons of efficiency and perhaps because of expectations about what knowledge is – that is, universal – wants the methodology to be widely applicable. However, the process of generalization results in a transformation that may make the revised methodology unsuitable to the original project – and proves liable to break the fragile connections that are supposed to unify the CDM system. In order to re-establish the missing link, the methodology needs to be changed again, or the project may not be enlisted. In turn, this requires a new involvement of the UN for approval, and each of these extra steps results in the project being slightly more delayed. The consequences of such delays were explained by another project developer:

> At the beginning of December we were ready, pretty much we were told that our project had been validated and that it needed to be signed off ... then we got an email a few days later saying sorry there's been a delay, the UN handed out a new manual for validation and so we've got a new set of standards by which we have to validate a project and so we're going to have to completely

---

20 As a result, some types of projects are known to dominate the CDM overall; for further discussion, see Giddens, op. cit., n. 12, p. 190.

re-start the validation of your project. And having been you know a few days away from getting the project registered we're now almost three months later and it still hasn't been registered, still hasn't been submitted for registration. When you think that ... these projects at least are worth about 3,000 Euros a day it can ... and also the fact that you're building a plant that's needed to supply energy onto the grid and you know the country is suffering from energy shortages while the validator is held up in an administrative nightmare.

But the CDM network is also one that becomes performed repeatedly, and developers soon become aware of which technologies, and which kinds of projects, will most easily be enlisted in the UN actor-world, and which projects they should therefore invest in. As the Chairman of a finance house explains:

> ... there's two main categories of [project] risk, there's the underlying risk of the project which influences how many emission reductions will actually happen and then there's the CDM risk which is the risk will the project actually get registered, will the reduction actually be verified and will the credits be issued. And there's risks as to whether or not those will happen, there's risks as to the quantity and there's risks as to the timing ...

Once registered, the operations of specific projects are themselves affected by the methodology with which they have become associated, and the enrolment of particular technologies in a project will have an impact on the ease with which it is run and set up. The actor-world is multiple, a quality which is determinant to understanding its key features, determinant of the progressive processes of enlisting of new actors, and also problematic in that it can result in restrictions or exclusions not anticipated. As an environmental consultant explained, some power projects are 'very simple'. A clear baseline of pre-CDM emissions can be established. Once technological changes have been made then it is often only a matter of reading a meter to see how much power and, by inference, how much less greenhouse gas is produced. Thus, some technologies lend themselves to the particular form of knowledge system developed within the CDM networks and projected by the UN. Those projects which have attempted to reduce emissions from diffuse sources (for example, households) are more difficult and therefore expensive to monitor. For example, once the UN agreed 'a formula' for the emission reductions that are achieved by distributing low-energy light bulbs, so monitoring costs fell by 25 per cent for one project developer (interview with a Senior Policy Manager of a social enterprise). But the system makes it more difficult for these projects to become enlisted, and crucially, excludes many diffuse solutions that integrate with the day-to-day routines and practices of, for example, local populations in developing states.

Individual projects can also become more likely to crumble when the accuracy of their measurements needs to be at their highest, which is the case with sites which emit higher levels of $CO_2$, as explained by a consultant to the off-shore oil industry:

So if we have to put in a new meter, we might close the [offshore] platform for a week. There are rules such as the meter has to be on straight bit of pipe in order that the meter is not affected by turbulence, so there are requirements about where the meter has to go. So, we might have to re-route pipe work. Then the meters must be inspected, calibrated in order that we can demonstrate that it meets the necessary tier of uncertainty. This job is required to be done by an independent company. If a meter fails, emissions will be guessed, and government approval must be sought for a plan to do this and how the meter is to be fixed.

This sort of complexity in turn necessitates an industry dedicated to financing, designing, validating, running, monitoring, verifying, and trading, and therefore very specific types of actors. These are service activities for which some countries, such as the United Kingdom, have an established set of agents and expertise. At the same time, the grounding of these practices in a very specific form of knowledge also means that those who do not own the relevant expertise will find it challenging to operate in the CDM networks – be it in relation to technologies, legal regulations or accountancy, for example. In sum, the particular emphasis put by the CDM on calculation and the importance of emission reduction credit allocation, participates in significant processes of exclusion within the CDM networks as designed by the UN. Only a particular form of scientific and technical knowledge is valued, which in turn means that only some technologies will be able to be enrolled into the CDM networks. In turn, the professional agents that become part of the process will be heavily co-dependent on the selected technologies. Our data thus support the argument of Lohmann that:

> the material nature of the carbon accounting discipline as well as the institutions that practice it has so far ensured that carbon credits flow to well-financed, high polluting operations capable of hiring professional validators of counterfactual scenarios, but not to non-professional actors eager to preserve or extend low-emitting livelihood practices or social movements actively working to reduce use of fossil fuels.[21]

## 2. Corruption

The failure of the CDM as originally envisaged goes even further, with issues that are a more substantial challenge to the nature of the actor-world as designed within the UN. In this section conflicts of interest are discussed which reportedly impact on the processes of validation and auditing, and are portrayed as absolutely critical to the credibility and financial worth of CERs. These processes are in part determined by a governance structure the nature and effectiveness of which is debated in the British media[22] and, as one might expect, by those we interviewed.

---

21 Lohmann, op. cit. n. 2, p. 513.
22 F. Carus, 'British deal to preserve Liberia's forests "could have bankrupted" nation' *Guardian*, 24 July 2010, 16; F. Hawkins, 'Liberia's inquiry into a carbon offsetting deal is a vital step forward' *Guardian*, 12 August 2010, 27.

An environmental consultant raised a fundamental criticism of the way in which the certification process is structured. This is that the DOEs who validate projects are employed and paid by project developers. This sets up what an environmental consultant terms an 'asymmetry of interest', and demonstrates how the concordance of interests of some powerful players in the CDM networks might influence the processes:

> ... if you look at financial audit you're the auditor, if you overstate the numbers someone will sue you and if you understate the numbers someone will sue you. If you don't get it right there's someone out there who's going to take you out of business ... In carbon audit if you overstate the numbers the project developer is happy, our buyer is happy, so everyone is happy. The people that aren't happy are the people from the next generation and the trees and animals, right. Well they're not going to sue you because they can't speak English, or they're not born or they have four legs.

Although the CDM system is supposed to be centrally focused on environmental interests, the co-generative power of some of the key actors can result in these sets of interests being sidelined. This specific aspect of the governance system was echoed by other interviewees. For example, although another consultant describes this situation in terms of its effects on those employed by DOEs, arguably the fault, if there is one, is with the system which specifies how and by whom validation is to be done.

> ... the DOEs are appointed by the project developers themselves (...) So if I'm a company I appoint a DOE, so the DOE has to certify my project and only then I will pay him ... on one side he has to comply with the UNFCCC procedures and guidelines, on the other hand the fee being paid to him is by the company, that is me, so he has to do a balancing act. And if he says no to a project then nobody else would go to that DOE further and he will lose business. (...) So it is a very tricky thing that the one who is validating the project is being paid by the same project developer.

This is a significant problem for a system which is portrayed as part of the international community's strategy to tackle climate change, and has been significantly scrutinized by the media. The outcome of this has been a challenge by various publics to the validity of the entire network suggested by the UN and, in particular, heavy questioning of whether it has the ability to fulfill any of the aims it set out to achieve.[23]

## RETHINKING THE ACTOR-WORLD

The 'testing' of assumptions of the CDM system put into place by the UN has not only led to a questioning of its value and operation. It has also brought practitioners such as development consultants to propose alternative visions of what this socio-legal technology may be about. In the process,

---

23 id.

they offer alternate visions of what this internationally supported mechanism may involve, and of the social models it may adopt.

One of the most controversial aspects of the CDM has been its link to sustainable development and, in particular, whether it was likely to meet its aims in this respect.[24] This has been a core aspect of the critiques directed at the CDM and its networks. Our data suggest that fundamental to the forms of response to the CDM with respect to sustainable development has been that some practitioners (as identified below) involved in the CDM propose an alternative social vision, or alternative 'actor-world', of what the CDM should be, what sort of society it should be part of, and in fact how 'sustainability' should be conceived. This has led to the emergence of forms of resistance to the world suggested and envisioned by the UN, and to the proposal of alternative visions. In contrast though, the malleability of the concept of sustainability was highlighted in several of our interviews.

As part of the approval process that a PDD needs to go through, the national authorities of the host country need to confirm that it will contribute to 'sustainable development' – therefore meeting one of the aims of the CDM. However, the concept of sustainable development is by no means self-explanatory, or one-dimensional:

> ... we rely on them [the DNA of a host country] as makers of policy and implementers of policy, environmental policy in that country to make that statement and stand behind us ... [but] ... I mean the phrase sustainable development really means nothing ... As a concept you know it's almost impossible to prove it to be honest. [project developer]

> This project design document has to be produced to the local government ... they check whether all the consents were established, consents to operate are up to date and available by the project developer ... But they really don't go into the real details of how many people were being employed or what is the other social community benefits from these projects. Of course they have to mention them but they really are only interested in the mandatory clearances yeah. [environmental consultant]

The fact that societal aspects of 'sustainable development' are thought to be excluded from the CDM networks is very meaningful in relation to the type of actor-world that was put forward by the UN, and the alternative visions that are offered. The latter, which serve as points of resistance to the UN actor-world, have been promoted alongside a well-known critique of the CDM, namely, that many of the projects that are currently part of the CDM system are also considered to be of little added value, as they would be taking place regardless of the mechanism.

Broadly speaking, there are three different forms of resistance. First, there has been an attempt to encourage more investment in CDM projects which

---

24 See, for example, A.G. Bumpus and J.C. Cole, 'How can the current CDM deliver sustainable development?' (2010) 1 *Wiley Interdisciplinary Revs.: Climate Change* 541.

are closer in form to a 'radical' vision of sustainable development.[25] By way of illustration, 42 different NGOs came together to articulate a set of standards for CDM projects known as the 'Gold Standard'.[26] These apply not only to the CDM but voluntary emission reductions schemes. As an environmental consultant said, conformity with the Gold Standard depends on a buyer who has motives other than 'reasons of compliance'. For most buyers in the CDM market, 'A CER is a CER whether it is Gold Standard or not. You don't need a Gold Standard on it ...' [manager responsible for CDM projects for a bank]. This is reflected in the small number of CDM projects which are registered with the Gold Standard scheme. This model proposes to review aspects of the CDM from within, and is designed to improve the system as it currently runs. Particularly central in this alternative model is the increased responsibility of the professions involved in the CDM.[27] The Gold Standard relies on an industry where striving for a 'true' engagement with 'sustainability' from the bottom up with the goal of less inequality should be central.[28] The revised actor-world is reliant on the industry moving in this direction, which has brought its own limitations and discrepancies.[29]

Second, there is a group active in the CDM who are self-described 'social entrepreneurs'. A defining characteristic of their projects is engagement with a large number of people in a given community to reduce demand for energy. For example, one project involves handing out low-energy light bulbs in Mexico [interview with a Senior Policy Officer of a social enterprise]. These projects are more costly in terms of measuring the emissions reduction achieved: in the past, methodologies have been approved which involved monitoring a sample of those households involved in the project, although more recently it has been agreed that assumptions can be made about emissions reduction from each bulb distributed. These projects are not full manifestations of radical sustainable development. A truly radical project would seek to alter the status quo in terms of equity, that is, by increasing

---

25 Redclift and Woodgate, op. cit., n. 9.
26 MacKenzie, op. cit., n. 2.
27 For a related engagement with the notion of 'ethics' in the CDM, see E. Boyd and G. Goodman, 'Tripping the Carbon Fantastic: The Clean Development Mechanism as Ethical Development?' (2011) EPD working paper no. 37, at <http://www.kcl.ac.uk/sspp/departments/geography/research/epd/workingpapers.aspx>; D. Liverman and E. Boyd, 'CDM development and ethics' in *New Mechanisms for Sustainable Development*, eds. K.H. Olsen and J. Fennhann (2008).
28 We are primarily interested here in analysing the proposal for a transforming aspects of the CDM put forward by practice; however, a reform of the governance system surrounding the CDM has also been called for in ways that cannot be isolated by academics and commentators; for example, see C. Sterck, 'The Clean Development Mechanism: the case for strength and stability' (2007) 2 *Environmental Liability* 91; E. Boyd et al., 'Reforming the CDM for sustainable development: lessons learned and policy futures' (2009) 12 *Environmental Sci. and Policy* 820.
29 See Bumpus and Cole, op. cit., n. 24; Boyd et al., id.

access to fuel or by working to place more control of a CDM project in the hands of local communities. Instead, the new vision offered for the CDM suggests enlisting more actors and transforming daily practices rather than industrial standards.[30] This implies an approach to tackling climate change potentially less heavily reliant on high-end technologies, and more closely linked to individual practices.

Third, there have been moves to produce of model whereby a 'social business' set up to alleviate poverty might, in part, be funded through the CDM. A definition of the social business model is given in a study of the feasibility of financing an Indian reforestation project via the CDM:

> At the core of the opportunity for poor land-owners is the social business model developed by the study ... By giving ownership of the business and contracting in the management and supervision capacity to operate on a large scale and in a manner that is credible in an industrialised market like forestry; the structure is designed to remove the barriers that prevent the poor trading in globalised markets ... The barriers to the poor benefiting include the capacities to: design effective industrial scale business projects, generate economies of scale, be dependable suppliers to industry, access financial support through grants, soft loans and commercial loans.[31]

So far, the 'barriers' identified by the report have yet to be overcome in practice. Most of the work conducted to date has concentrated on piloting the process to bring together many small foresters and farmers to agree on the aims and management structures required to develop and run a CDM project. But it is interesting to note how this approach goes a step further in challenging the actor-world put forward by the UN. Here, the political-economic challenge is at its clearest, as the alternative version of the mechanism becomes not only more 'encompassing', but also a tool to bridge some of the growing gaps between rich and poor.

With all three examples, we see how alternative visions for the global socio-legal tool that is the CDM, can be deployed as a response to the failures and limitations of the dominant model. Each of these is likely to meet its own set of challenges, and remains open to difficulties and failure if the elements that need enlisting fail to be mobilized – be they more conscientious practitioners, technologies that can be fed within the UN system, or global support for a new political vision for carbon market mechanisms.

---

30 For an analysis of the links between climate change policies and daily practices, see P. Macnaghten, 'Embodying the environment in everyday life practices' (2003) 51 *Sociological Rev.* 63.
31 N. Pyatt et al., *Sustainable Industrial Sequestration Planting Feasibility Study* (2009).

## CONCLUSION

The form of development that is inherently inscribed into the CDM and some of its dominant features are expressed and materialized in the practical workings of the CDM. The embedded 'scripts', or sets of ideas and values, are only realized when the carbon market operates, raising concerns for those who created it when confronted with forms of failure or transformation. The future as envisaged is resisted and rewritten via processes of exclusion and corruption, not only from without but also from within.

This volume sets out to explore the materiality of science, technology, and law, and this article analyses the CDM as an actor-world to reveal something of the effort which goes into turning a particular vision into a stabilized mechanism allowing for individual technological transformations to operate, whilst limiting the opportunities for certain others to happen. In this process of translation, the legal text came not only to embed a particular script, but also to articulate the series of rules and principles that are expected to be followed if the translation is to take place in the way the makers of the Kyoto Protocol expected it to. The process of translation from idea to CERs that any project goes through is mediated by legal transactions and travels through lawyers' offices, from decisions made at Kyoto and embedded in the protocol, to the series of contracts and legal transactions that make a project feasible, and the series of legal transactions involved in the issuing, buying, and selling of CERs. Our findings suggest that understandings of the law might be multivalent and contested with regard to climate change and the CDM in particular. Whilst further research on how various understandings are articulated in practice, and how individual legal transactions come to participate in the deployment of particular networked effects, would certainly be useful, it is interesting to reflect on other dimensions of the law here, in particular, the role of the legal as the global framework through which the CDM was created.

The first element that our analysis points to is a reimagining of the classic 'gap' issue,[32] a common feature of socio-legal analyses, when the law, envisaged as series of relatively straightforward and causal links, encounters failures and transformations. However, an account influenced by the concept of actor-worlds suggests that underlying conflict of vision is almost inevitable, necessitating a political-economic rather than merely technical response, whether in terms of law or material transfers.

The second element highlights various strategies of response and resistance to the actor-world that was articulated in Kyoto. Administrator-sociologists deploy several approaches to the practical enactment of their own critique of the CDM, and of the actor-world it embeds. Their relationship with the law, in that context, is interesting to reflect upon. The methods

---

32  R. Pound, 'Law in Books and Law in Action' (1910) 44 *Am. Law Rev.* 12.

deployed are worked within the legal framework, and offer a transformation of the mechanism from within rather than a direct confrontation with it. The CDM is challenged but remains engaged with. Its materialization through the law has made it an obligatory passage[33] point that actors need to engage with in order to support the projects that they see as central to the writing of the newly sustainable actor-world they are hoping to build. The formal framework of the CDM, and the carbon market more generally, is being challenged by others elsewhere, but there is a notable attempt to engage with the law and its embedded value that involves a rewriting from within which those administrator-sociologists who may not wish or be in a position to engage at other levels, are deploying. Making an observation on the Gold Standard, Donald MacKenzie notes:

> The intervention by the World Wildlife Fund and other NGOs was informal: it did not alter the formal procedures of the CDM. However, NGOs are also seeking to practise a politics of market design in a more formal sense, seeking to alter rules and procedures. That, indeed, is precisely the course of action that Callon and Latour's perspective implies. If markets are plural – Callon's best-known work is titled *The Laws of the Markets* (Callon, 1998) – and 'capitalism' has no unalterable essence, then this may indeed be productive.[34]

Our analysis shows that actors engaging with alternative modes of resistance propose new ways of reading and enacting the law. This in turn suggests that they are challenging the stabilized vision in contemporary international law of what 'environmental sustainability' means.

---

33 As developed in M. Callon, 'Some Elements of a Sociology of Translation: Domestication of the Scallops and the Fishermen of St Brieuc Bay' in *Power, Action and Belief: A New Sociology of Knowledge?*, ed. J. Law (1986) 196–233.
34 MacKenzie, op. cit., n. 2, p. 452.

# Nanotechnology and the Products of Inherited Regulation

ELEN STOKES*

*New technologies do not always elicit new regulatory responses. More often than not, policymakers deal with new technologies by deferring to existing regulatory regimes. This article argues that there are often overlooked consequences of grafting a new technological area, displaying different types of risks and uncertainties, onto an existing regulatory framework. Not only can it entail the application of ill-suited rules and standards, but it can also involve the reproduction of deeply ingrained traditions and assumptions which, under the weight of history, makes scrutiny extremely difficult. As is shown here, nanotechnology-enhanced products inherit a raft of consumer protection rules as well as a regulatory predisposition to internal market facilitation. So entrenched is the focus on market opening that making ad hoc changes to existing regulations to incorporate the broader concerns around nano-products cannot escape the reach of a very powerful market context.*

## INTRODUCTION

This article looks at the consequences of applying old regulatory measures to new regulatory problems. It addresses, along with the volume as a whole, what happens when legal regulation confronts and adapts to increasingly complex and contested technological worlds. Much has been written about the inability of the law to keep up with technological innovation in the

---

* Cardiff Law School, Law Building, Museum Avenue, Cardiff CF10 3X, Wales, and ESRC Research Centre for Business Relationships, Accountability, Sustainability and Society (BRASS), Cardiff University, 55 Park Place, Cardiff CF10 3AX, Wales
StokesER@cardiff.ac.uk

I am grateful to Liz Fisher, Antonia Layard, Bob Lee, Marie Lee, and Steven Vaughan for helpful comments on an earlier draft. I retain sole responsibility for the content, including any errors or omissions.

marketplace.[1] The relationship is typically characterized as one of initial struggle, a race between science's hare and law's tortoise, in which the law marches ahead 'but in the rear and limping a little'.[2] A commonly held perception is that, unless and until the law responds, the latest high-tech products will enjoy an initial period without regulation.

Yet, the reality is often quite different. It is hard to imagine that a new technology could ever be completely 'lawless'.[3] Given the vast number of EU measures protecting occupational health and safety, public health, and the environment, which have wide and overlapping remits, it is inconceivable that the latest high-tech products could slip through the regulatory net. Most of the time there is already legislation in place to deal with unwanted outcomes. Where these provisions impose general standards such as 'safety' and 'risk prevention', they are considered to be broad enough to cover all technological developments, even those that were unimaginable when the rules were drafted.[4]

However, regulatory coverage is no guarantee of regulatory adequacy, for it may well be the case that the measures relied upon are ill-suited to the threats and opportunities of technological progress. This is one of the more troublesome and less well understood implications of the reverence shown for prior regulations. As is shown here, the use of existing provisions to regulate a new area carries its own operational and ideological baggage. In practice it involves more than the application of old rules to new products; it entails the continued presence of the policy substructure, the underlying aims and assumptions of those rules.

To illustrate, the article looks at the emergence of new consumer products containing nano-materials. Section I introduces 'nanotechnology' before exploring the politics of its regulation. It explains that, in spite of disagreement between EU institutions, the Commission's view is that nano-products can be regulated using existing measures designed to deal with conventional types of product. Section II suggests that the extrapolation of specific rules from conventional products to functionally different nano-products is problematic, because it cannot be assumed that the same standards or assessment requirements should be equally applied. The difficulties run much deeper than this, however, where nano-products inherit not only regulatory rules but regulatory predispositions. Consumer protection regulation in the EU seeks, above all else, to facilitate the functioning of the internal market. Since nano-products are presumed to fall within the

---

1 See, for example, R.A. Posner, *Catastrophe: Risk and Response* (2004) 8; G.E. Marchant, 'The Growing Gap Between Emerging Technologies and the Law' (2011) 7 *International Library of Ethics, Law and Technology* 19–33.
2 Windeyer J in *Mount Isa* v. *Pusey* (1970) 125 CLR 383, at 395.
3 G.V. Calster, 'Regulating Nanotechnology in the European Union' (2006) *Nanotechnology Law & Business* 359–372, at 360.
4 D. Friedman, 'Does Technology Require New Law?' (2001) 25 *Harvard J. of Law & Public Policy* 71–85.

same policy remit as conventional products, their regulation is similarly market-oriented. So far, the policy debate has centred on preventing possible risks by limiting the availability of nano-products until more is known about their impacts. Yet, as this article shows, the current regulatory framework is already programmed to open up, not close down, markets.

The emergence of new, technologically enhanced products into an inherited regulatory environment ought to provide an opportunity for re-examining the current approach. However the presumed elasticity of regulations forecloses full and proper debate about the suitability and starting points of the regulatory regime. Moreover, it makes questioning the desirability of new, innovative products difficult since their introduction into the market becomes an inevitable and incontrovertible consequence of their supposed regulation. The fact that a regulatory net is in place gives the green light to product circulation, and there is no pause to consider the repercussions, regulatory or otherwise, of their widespread availability.

Section III suggests that one solution might be to amend existing legislation to deal specifically with the concerns around the technology in question. This approach is currently being adopted in relation to specific types of nano-product. For example, nano-labelling requirements have already been inserted into regulations on cosmetic products and similar provisions are being considered in respect of foodstuffs, biocidal products, and electrical equipment. Yet, making ad hoc changes to existing regulations does not escape the reach of the very powerful internal market context. It is shown here that, although these alterations seek to achieve a more suitable response to nano-products, they may nonetheless feel the constraining effects of their regulatory and policy predecessors.

## I. REGULATORY CHALLENGES

*1. Nanotechnologies: the shape (and size) of things to come*

Some of us, perhaps without even knowing it, will have bought products containing nano-materials. Nano-products have been marketed for some time and their availability looks set to increase dramatically. The inclusion of nano-materials, deliberately engineered to achieve specific properties, presents new opportunities to increase the performance of traditional products and develop unique new product lines. One of the key determinants of improved functionality is size. Generally speaking, materials with one or more dimension at the nano-scale (1–100 nanometres, one nanometre being one billionth of a metre[5]) can exhibit unusual characteristics not shared by the same material in its conventional, bulk form. For instance, materials that

---

5 British Standards Institute (BSI), *Vocabulary – Nanoparticles*, Publicly Available Specification 71 (2005) 2.

are opaque in their conventional form can become transparent in nano form; likewise those that are insulators at bulk scale can become highly effective conductors when nano-size; inert substances can become catalytic; and brittle materials can show signs of improved strength and resistance to fracturing.[6]

These novel properties are already being exploited, with a range of applications spilling out of the laboratory and into the market. One recent estimate suggests that there are over 1,300 nano-based consumer products available for purchase.[7] Odour-free socks, for example, tend to be coated in nano-scale silver particles for their anti-bacterial, anti-microbial properties. Stain-resistant garments and upholstery contain nano-whiskers that help to repel liquids. Non-stick cookware often has nano-composite surfaces. Many sunscreens are made using nano-scale titanium dioxide or zinc oxide because they offer enhanced protection against UV rays, and, unlike conventional lotions, they are clear, not white. Nanotechnology is believed to offer lots of other extraordinary economic and societal benefits, as evidenced by the significant and sustained global investment in research and development.[8] There are growing concerns, however, that the very characteristics that make nano-products commercially attractive will also lead to new risks to human health and the environment.

One reason for this is that nano-materials have extremely small dimensions, which means that they have a proportionally larger surface area per unit mass than the same materials in their bulk form. This can lead to an increase in reactivity (because more surface is available to react with other substances) and hence an increase in potential toxicity. Not all nano-materials are associated with exaggerated risk profiles, with many posing little or no added threat. The immediate concern is with engineered (rather than naturally occurring) nano-materials and those in the form of free particles, such as in powders or liquids (rather than fixed in a solid matrix), although their toxicology is by no means fully understood. The problem is that whilst we can begin to hypothesize about the unusual toxicity properties of carbon nano-tubes,[9] for instance, or the capacity of some nano-particles to cause DNA damage across cellular walls,[10] we are not yet in a position to articulate risks precisely.

---

6 For further detail, see M. Ratner and D. Ratner, *Nanotechnology: A Gentle Introduction to the Next Big Idea* (2003).
7 Woodrow Wilson International Center for Scholars, Nanotechnology Consumer Products Inventory, at <http://www.nanotechproject.org/inventories/consumer/>.
8 Cientifica, *Global Nanotechnology Funding 2011* (2011).
9 See, for example, K. Donaldson and C.A. Poland, 'Nanotoxicology: New Insights into Nanotubes' (2009) 4 *Nature Nanotechnology* 708–10; K. Donaldson et al., 'Review: Carbon Nanotubes: A Review of Their Properties in Relation to Pulmonary Toxicology and Workplace Safety' (2006) 92 *Toxicological Sciences* 5–22.
10 G. Bhabra et al., 'Nanoparticles can cause DNA damage across a cellular barrier' (2009) 4 *Nature Nanotechnology* 876–83.

This goes to show that, in spite of its initial commercial success, nanotechnology is still in its infancy and more information is needed on its implications. The area is plagued by uncertainty relating to aspects as varied as its promises, expectations, technical feasibility, social acceptability, and ethical desirability, although for the most part the focus is on the unknown parameters of the physical harm likely to ensue.[11] Whilst there is some evidence to suggest that certain materials at the nano-scale are potentially more hazardous than their conventional bulk-scale counterparts, little is known about how this translates into risk (risk being a product of magnitude and likelihood of impact in a given exposure scenario). This poses the greatest problem for policymakers. Not only are they confronted with extensive knowledge gaps but they also need the tools to identify and deal with the unknowns. There is a current lack of validated measurement methods and detection techniques,[12] and there is no widely accepted systematic approach for evaluating potential consumer risks, even though nano-materials can be found in a number of commercially available products.[13] Whilst this is not unusual for a fledgling technology like nanotechnology, it is no doubt compounded by the rapid development of new nano-products which currently outstrips our capacity to test plausible impacts. Finding a solution has been recognized as a matter of priority, particularly in the EU where demands for clarity have grown in number and vigour.

## 2. *The big politics of small science*

Persistent uncertainties along with concerns about the possible consequences of human and environmental exposure has fuelled fierce debate as to how nano-materials are best managed. Some stakeholders have made repeated calls for temporary bans on all nano-materials,[14] on certain types, or on their inclusion in certain types of product[15] pending more robust investigations

---

11  Others note that the focus on risk and safety is a narrow one: see M. Lee, 'Risk and Beyond: EU Regulation of Nanotechnology' (2010) 6 *European Law Rev.* 799–821; C. Groves, 'Nanotechnology, Contingency and Finitude' (2009) 3 *NanoEthics* 1–16; R.G. Lee, 'Look at Mother Nature on the Run in the Twenty-First Century: Responsibility, Research and Innovation' (2012) *Transnational Environmental Law* (forthcoming).
12  L.C. Abbott and A.D. Maynard, 'Exposure Assessment Approaches for Engineered Nanomaterials' (2010) 30 *Risk Analysis* 1634–44.
13  T. Thomas et al., 'Moving Toward Exposure and Risk Evaluation of Nanomaterials: Challenges and Future Directions' (2009) 1 *Wiley Interdisciplinary Revs.: Nanomedicine and Nanobiotechnology* 426–33, at 427.
14  ETC Group, 'Nanotech Product Recall Underscores Need for Nanotech Moratorium: Is the Magic Gone?' press release, 7 April 2006.
15  European Parliament (EP), Report on the Proposal for a Directive on the Restriction of the Use of Certain Hazardous Substances in Electrical and Electronic Equipment (Recast), 15 June 2010, A7-0196/2010, Amendment 88; EP Debate No. 9, 10 July 2010, Explanation of Votes, MEP Robert Rochefort; for examples outside the EU,

into their likely behaviour or 'the certainty of every risk having been removed'.[16] The European Parliament – specifically its Committee on the Environment, Public Health and Food Safety – has waged a lengthy campaign for the amendment of Community legislation specifically to address nano-materials and the products that contain them.[17] The concern is that, because there is no separate provision for nano-products, current regulations fail to provide safeguards against possible harm, as one Member of the European Parliament explains: '[w]e cannot simply allow these products to be put onto the market and tested on consumers; we cannot allow consumers to be treated as guinea pigs'.[18] A more proactive approach is needed, because '[a]t the moment, we are stepping on the gas of nanotechnology without first ensuring that we have emergency brakes or even knowing whether the steering is working.'[19]

Others actors and institutions, by contrast, are critical of the idea that regulations ought to be updated to account for nano-materials,[20] especially where it would involve reacting 'against imagined risks, merely because they are in something that is so small as to be difficult to identify, or even, dare I say, to understand'.[21] Whereas the European Parliament has been an enthusiastic supporter of change, the Commission has cast itself as gatekeeper of the status quo. Current legislation, says the Commission, is already equipped to deal with nano-products even if it does not directly refer to 'nano-materials', 'nanotechnology' or the 'nano-scale'.[22] Until recently, this has remained the dominant position in the EU, reinforcing the notion that the regulatory future of nanotechnology is to be governed by its regulatory past.

see International Center for Technology Assessment, Citizen Petition to the United States Food and Drug Administration: Petition Requesting FDA Amend its Regulations for Products Composed of Engineered Nanoparticles Generally and Sunscreen Drug Products Composed of Engineered Nanoparticles Specifically, filed 17 May 2006; Australian Education Union Victoria, Council Resolution adopted 13 May 2011, at <http://www.aeuvic.asn.au/80284.html>.
16 EP Debate No. 4, 28 September 2006, MEP Hiltrud Breyer.
17 EP Resolution on Regulatory Aspects of Nanomaterials, P6_TA(2009)0328.
18 EP Debate, op. cit., n. 16.
19 id., MEP David Hammerstein Mintz.
20 See, for example, Commission, Classification, Labelling and Packaging of Nanomaterials in REACH and CLP, CA/90/2009/Rev 2.
21 EP Debate, op. cit., n. 16, MEP Giles Chichester.
22 Commission, Regulatory Aspects of Nanomaterials COM(2008) 366 final, 4, 8; also EP Debate, op. cit., n. 16, Commissioner Stavros Dimas.

## II. REGULATING IN THE SHADOW OF EXISTING LAW

*1. Inherited legislation*

The Commission's stance, characterized by its deference to existing law and policy, offers one illustration of inherited regulation. There are several explanations for the continued application of measures designed with more conventional products in mind. First, the Commission has remained preoccupied with the question of regulatory coverage, focusing on whether nano-materials and/or their host products fall within the definitional remit of existing provisions. This line of reasoning is most salient in relation to products regulation because provisions are aimed at broad categories of goods (foods, cosmetics, detergents) without distinguishing between nano and conventional variants. Food comprising nano-material ingredients is still 'food' for the purposes of the General Food Regulation,[23] for instance, and will be subject to the legislative requirements regardless of its particle size and composition. The same is true for other products. For example, cosmetics attract legal obligations because they fall within the comprehensive definition ('substance or mixture'[24]) set out in the Cosmetic Products Regulation,[25] not because they comprise materials at any particular scale. Even where provisions are directed at regulating chemicals that are later incorporated into products, a similarly broad-brush approach is adopted. Here the Commission notes that although nano-materials are not referred to in EU chemicals legislation, they nonetheless fall within the definition of 'substance' which triggers legal obligations.[26] Elsewhere it explains that:

> [w]here regulation contains requirements of a general nature, they will cover also risks related to nanotechnology, even if they have been adopted without specifically intending to address risks associated with nanomaterials and nanotechnologies.[27]

This definitional inclusiveness explains, at least in part, the Commission's finding that current frameworks are broad enough to capture nano-products.

---

23 Regulation (EC) No. 178/2002 laying down the general principles and requirements of food law, establishing the European Food Safety Authority, and laying down procedures in matters of food safety [2002] OJ L31/1, Article 2.
24 Regulation (EC) No. 1223/2009 on Cosmetic Products, [2009] OJ L342/59, Article 2(1)(a).
25 id.
26 Commission, op. cit., n. 20; also see Regulation (EC) No. 1907/2006 concerning the Registration, Evaluation, Authorisation and Restriction of Chemicals (REACH) [2006] OJ :396/1: Article 1(3) imposes a duty of care on manufacturers and suppliers to ensure that the substances they manufacture, supply or use do not adversely affect human health or the environment.
27 Commission, Commission Staff Working Document: Accompanying Document to the Communication on 'Regulatory Aspects of Nanomaterials' SEC(2008) 2036, 12.

A second explanation lies in the EU's already extensive programme of consumer protection legislation. The first signs of coordinated Community level measures on the consumer emerged in the early 1970s,[28] propelled by policy moves towards the progressive harmonization of economic law which later evolved into the internal market. At the time, specific references to consumers were few and far between, arising only in the contexts of agricultural policy, which guaranteed availability of supplies and the stabilization of markets, and the rules of competition, which sought to limit the incidence of unfair practices. The pace of legislative activity soon gathered under pressure from the European Parliament[29] resulting in the adoption of a preliminary programme – the first of its kind in the Community – for a consumer protection and information policy.[30] Since then the policy sphere has become densely populated with measures such as those addressing the quality and price of products,[31] measures redressing the imbalance of power between producers and consumers,[32] measures setting standards of safety, and measures requiring the disclosure of information relevant to consumer transactions. Whilst, over time, the parameters of consumer policy have steadily expanded,[33] core activities have continued to centre on three principal strands: the economic and legal protection of consumers; the physical protection of consumers; and consumer information and education.[34]

This elaborate system of rules and principles applies in relation to all products intended for or made available to consumers, irrespective of their composition or characteristics. Consequently nano-products are required to meet certain standards for the simple reason that they are traded in the EU. Whether, and to what extent, the specific content of legislation is appropriate has been questioned elsewhere.[35] One concern is that standards requiring certain conditions of quality to be met at the point of supply, or those compelling the manufacturer to use certain production methods or materials,

28 Commission, Decision Relating to the Setting Up of a Consumers' Consultative Committee 73/306/EEC [1973] OJ L283/18.
29 Commission, Consumer Protection and Information Policy: First Report (1977) para. 11.
30 Council, Resolution on a Preliminary Programme of the European Economic Community for a Consumer Protection and Information Policy [1975] OJ C92/2.
31 For example, Council Directive 74/409/EEC on the harmonization of the laws of the Member States relating to honey [1974] OJ L221/10; Council Directive 75/726/EEC on the approximation of the laws of the Member States concerning fruit juices and certain similar products [1975] OJ L311/40.
32 Council, op. cit., n. 30, paras. 6–12.
33 For a useful overview, see S. Weatherill, *EU Consumer Law and Policy* (2005) ch. 1.
34 Commission, Consumer Protection and Information Policy: Second Report (1979) 9.
35 See, for example, Royal Society/Royal Academy of Engineering, *Nanoscience and Nanotechnologies: Opportunities and Uncertainties* (2004) 69; L. Frater et al., *An Overview of the Framework of Current Regulation affecting the Development and Marketing of Nanomaterials: A Report for the DTI* (2006); E. Stokes, 'Regulating Nanotechnologies: Sizing Up the Options' (2009) 29 *Legal Studies* 281–304.

are tailored to conventional types of product and may be ill-equipped to deal with the peculiarities of nano-scale equivalents.[36] The purpose of this article is not to revisit these arguments but, rather, to show that the difficulty runs deeper than regulatory content. This is because the presumed application of existing measures to nano-products entails more than the replication of regulatory requirements. It also involves the transmission of traditions and assumptions, inbuilt in the regulatory regime, about how those requirements should apply and to what ends. In this sense, consumer protection legislation not only provides the particularities of nano-products regulation, such as specific codes or standards, but it imposes its own framework of regulatory values, goals, intentions, priorities, and powers as well as a range of interpretative practices. When nano-products are caught within the definitional remit of a particular Directive or Regulation, therefore, they inherit the scope and content of those provisions but also the broader policy setting.

An exploration of the policy subtext reveals the assumptions and defining pressures that have shaped the genesis and form of regulations now applied in spheres of consumer protection. Consumer policy in the EU exerts pressures in different directions as it purports to address a kaleidoscopic range of issues. The Commission notes, for example, that policies ought to have multiple foci, 'it being agreed that it was not enough to deal solely with the market place'.[37] Community policy apparently has other interests in mind, such as the 'everyday reality in the lives of its citizens'[38] and 'the hearts of men and women and not merely with the management of packages'.[39] From the policy discourse emerges a European consumer that is 'no longer seen merely as a purchaser'[40] but one that 'is concerned with the various facets of society which might affect him directly or indirectly'.[41] Consumption is punctuated with questions and concerns arising beyond those involved in the simple exchange of money for products.[42] It is anything but the private and apolitical affair it is sometimes presented as being, and this is reflected in the policy undertaking to accommodate a diverse range of consumer needs and identities. The EU consumer is many things, but some of the most common policy projections are of a safe,[43]

---

36 Stokes, id., pp. 284–9.
37 Commission, op. cit., n. 29, para. 16.
38 id., p. 8.
39 id.
40 Council, op. cit., n. 30, para. 3.
41 id.
42 For discussion, see R. Sassatelli, 'Virtue, Responsibility and Consumer Choice: Framing Critical Consumerism' in *Consuming Cultures, Global Perspectives: Historical Trajectories, Transnational Exchanges*, eds. J. Brewer and F. Trentmann (2006) ch. 9; D. Stolle et al., 'Politics in the Supermarket: Political Consumerism as a Form of Political Participation' (2005) 26 *International Political Sci. Rev.* 245–69.
43 Commission, Product Safety: Ensuring the Safety of Products for All EU Citizens (2006); Commission, Your Rights as a Consumer: How the European Union Protects Your Interests (2007) 5–6.

confident,[44] informed,[45] demanding,[46] sovereign,[47] rational,[48] empowered,[49] and citizen-like[50] individual.

## 2. Consumers consume

Yet, in spite of the numerous policy constructions, the role of the EU consumer is ultimately to consume. Community consumer protection is driven, first and foremost, by the EU's 'core project',[51] the internal market. 'Market opening and consumer policy', it is said, 'go hand in hand'.[52] The first ever Product Liability Directive[53] aimed to 'unite consumers' interests with Single Market policies (namely, free exchange of goods and elimination of competition distortions)'.[54] The first General Product Safety Directive[55] (which covered 'any product intended for consumers or likely to be used by consumers'[56] including food) had similar motives, stating that 'it is important to adopt measures with the aim of progressively establishing the internal market'.[57] At the time, consumer protection was recognized as a policy goal in itself under the Single European Act,[58] although the Act did not offer a specific legal basis for consumer legislation. As a result, early

---

44 S. Weatherill, 'The Evolution of European Consumer Law and Policy: From Well Informed Consumer to Confident Consumer?' in *Rechtseinheit oder Rechtsvielfalt in Europa? Rolle und Funktion des Verbraucherrechts in der EG und den MOE-Staaten*, ed. H.W. Micklitz (1996) ch. 25.
45 Commission, Communication on the EU Consumer Policy Strategy 2007–2013 COM(2007) final 99, 22.
46 Commission, DG Health and Consumers Management Plan 2011 and Beyond (2010) 5.
47 M. Nardo et al., *The Consumer Empowerment Index* (2011) 15–16.
48 Commission, op. cit., n. 45, p. 10.
49 id., p. 13; M. Monti, *A New Strategy for the Single Market: At the Service of Europe's Economy and Society* (2010) 41.
50 Commission, Communication on Consumer Policy Action Plan 1999–2001 COM(98) 696 final, 1; for discussion, see M. Bevir and F. Trentmann, 'Civic Choices: Retrieving Perspectives on Rationality, Consumption, and Citizenship' in *Citizenship and Consumption*, eds. K. Soper and F. Trentmann (2007) ch. 1.
51 Commission, Green Paper on European Union Consumer Protection COM(2001) 531 final, 3.
52 Commission, Communication on a Single Market for 21st Century Europe COM(2007) 724 final, 4; also Commission, Follow-up Communication to the Green Paper on EU Consumer Protection COM(2002) 289 final, 3.
53 Council Directive 85/374/EEC concerning liability for defective products [1985] OJ L21/29.
54 Commission, Third Report on the Application of Council Directive on the Approximation of Laws, Regulations and Administrative Provisions of the Member States Concerning Liability for Defective Products COM(2006) 496 final, 4.
55 Council Directive 92/59/EEC on General Product Safety [1992] OJ L228/24.
56 id., Article 2(a).
57 id., Preamble.
58 SEA [1987] OJ L169/1.

consumer measures were adopted on the ground of Article 100a EC (later Article 95 EC, now Article 114 TFEU) because they 'have as their object the establishment and functioning of the internal market'.[59]

Even when legislative activity in the area was later supplied with a specific legal basis,[60] consumer measures have tended to retain their internal market roots. For instance, one reason for the introduction of the first Cosmetics Directive was that differences between member state laws could 'hinder trade in these products and, as a result, have a direct effect on the establishment and functioning of the common market'[61] and this was reflected in its legal basis.[62] Of course, some would argue that the real reason for such measures is the protection of consumer health which is pursued in tandem with free trade objectives.[63] However, the notion that divergences between national consumer laws impedes the functioning of the internal market retains a strong hold over EU jurisprudence, to the extent that Community competence to harmonize market conditions is, in practice if not formally, virtually unlimited.[64] Hence, the Commission, in recasting the Cosmetics Directive 23 years after it was first promulgated, noted that this rationale continues to be valid because 'Community action is necessary to avoid a fragmentation of the market and to ensure a high and equal level of protection of the European consumer'.[65] As a result, the same legal basis was used. Similar stories emerge from other sectors too. The power to legislate on low-voltage electrical equipment initially derived from the internal market legal basis;[66] when it came to be replaced, the same legal basis was adopted.[67] Early legislation on the safety of toys took the internal market as its basis;[68] so too did successive measures on toys.[69] Interestingly, whereas the first Community Directive on chemicals had two legal bases (internal

---

59 Treaty establishing the European Community [1992] OJ C224/6, Article 100(a)(1).
60 Treaty on European Union [1992] OJ C191/1, Title XI, Article 129a, replaced by the Treaty of Amsterdam Amending the Treaty on European Union, the Treaties Establishing the European Communities and Certain Related Acts [1997] OJ C340/173, Article 129a, now enshrined in Treaty on the Functioning of the European Union [2008] OJ C115/47, Title XV, Article 169.
61 Council Directive 76/768/EEC on Cosmetic Products [1976] OJ L262/169, Preamble.
62 Article 95 EC.
63 Case C-210/03, *Swedish Match* v. *Secretary of State for Health* [2004] ECR I-11893, para. 33.
64 For discussion, see S. Weatherill, 'The Limits of Legislative Harmonization Ten Years After *Tobacco Advertising*: How the Court's Case Law Has Become a "Drafting Guide"' (2011) 12 *German Law J.* 827–64.
65 Commission, Corrigendum: Proposal for a Regulation on Cosmetic Products (Recast) COM(2008) 49 final/2, 4.
66 Low Voltage Directive 73/23/EEC [1973] OJ L77/29, Preamble.
67 Low Voltage Directive 2006/95/EC [2006] OJ L374/10, Preamble.
68 Toy Safety Directive 88/378/EEC [1988] OJ L187/13, Preamble.
69 Toy Safety Directive 2009/48/EC [2009] OJ L170/1, Preamble.

market and 'residual powers'),[70] new chemicals legislation has kept only one: the internal market.[71] When the overarching General Product Safety Directive was updated,[72] it too was underpinned by internal market goals.

This legislation is engaged when products containing nano-materials are introduced to the market, either because the products are covered by sector-specific legislation in areas such as cosmetics or foodstuffs, or because the products are caught by the General Product Safety provisions. The legislative measures are already oriented in a particular direction, overwhelmingly towards market harmonization, and this provides the backdrop against which nanotechnologies now emerge. Naturally, the improved functioning of the single market is not the only legislative aim stated. Community consumer policy also seeks to guarantee 'effective protection and solid rights'.[73] It aims to achieve this by safeguarding against 'the serious risks and threats that they cannot tackle as individuals. A high level of protection against these threats is essential to consumer confidence'.[74] What is required is 'a sophisticated regulatory framework and enforcement measures to ensure that markets can operate in an environment of high levels of safety and consumer confidence'.[75] The emphasis on safe and confident consumers is to be expected in any strategy on protection. However, a sense of their capacity to secure market goals is never lost, and they are construed primarily in instrumental terms in that 'a high level of confidence in the safety of goods and services is *essential to the stability of markets and of trade* within the internal market in particular'.[76] Confident consumers are an integral part of the 'motor of economic evolution'.[77]

Building in both weight and momentum since the first phase of Community consumer action, these assumptions and inclinations have also come to frame the regulation of nanotechnologies. Even the earliest of initiatives on new technologies, such as telecommunication and computer technologies, was similarly geared towards the 'core project', as evidenced by the Commission's statement:

> The development of new technologies has led to the creation and development of new cross-border services which are playing an increasingly important role in the economy. However, these services can develop their full potential only when they serve a large, unobstructed market.[78]

---

70 Dangerous Substances Directive 67/548/EEC [1967] OJ L196/1, Preamble.
71 REACH Regulation, op. cit., n. 26.
72 Directive 2001/95/EC on General Product Safety [2002] OJ L11/4.
73 Commission, op. cit., n. 46, p. 8.
74 Commission, op. cit., n. 45, p. 13. This objective is repeated throughout the Treaty, see, for example, Articles 114(3), 168(1), 169(1) TFEU.
75 Commission, op. cit., n. 46, p. 3.
76 id., emphasis added.
77 Commission, op. cit., n. 45, p. 4.
78 Commission, White Paper on Completing the Internal Market COM(1985) 310 final, 32.

This legacy lives on through the policy statements on nanotechnology which, though emphasizing the importance of 'a high level of care for public health, safety, the environment and consumer protection',[79] are 'aimed at exploiting the huge potential of nanosciences and nanotechnologies in research and innovation'.[80] This is crucial if the EU is to gain any competitive advantage in the field, it is argued, since:

> [w]e Europeans have to be aware of the fact that we will not remain leaders in a range of markets and technologies for ever, and that there are many markets in which we have already lost it, and along with it the power to control many technologies.[81]

Likewise, there have been calls for the EU to 'coordinate, lead and hold the Member States to account'[82] so that industrial policy provides Europe with the opportunity to assume pole position in 'manufacturing industries, biotechnology, nanotechnology, the chemical industry'.[83] In order to achieve this, and to ensure that R&D is 'translated into commercially viable, inherently safe products and processes',[84] the policy focus is on 'actions paving the way to a level playing field for nanotechnology-based products in the globalised market'.[85] In line with existing frameworks of consumer protection, the central aim is harmonization, consistency, and the avoidance of market distortions.[86] However, the weight of history is felt in the additional sense that:

> Europe must avoid a repeat of the European 'paradox' witnessed for other technologies and transform its world-class R&D in N&N [nanosciences and nanotechnologies] into useful wealth-generating products in line with the actions for growth and jobs, as outlined in the 'Lisbon Strategy' of the Union.[87]

The fear is that, unless R&D is readily translated into products and profits, Europe will lose its competitive advantage. This sentiment is shared by others who call for 'a coherent and coordinated strategy for accelerating the application of nanotechnology as widely as possible across the economy'.[88] Hence the thrust of nanotechnology regulation, the primacy it affords to market liberalization and the use of consumption as a yardstick of market

---

79 EP Debate, op. cit, n. 16, Commissioner Janez Potoènik.
80 id.
81 id., MEP Jorgo Chatzimarkakis.
82 EP Debate No. 18, 8 March 2011, MEP Edit Herczog.
83 id.
84 Commission, Communication on Nanosciences and Nanotechnologies: An Action Plan for Europe 2005–2009 COM(2005) 243 final, 7.
85 id., p. 11.
86 Commission, Towards a European Strategy for Nanotechnology COM(2004) 338 final, 18.
87 Commission, op. cit., n. 84, p. 2.
88 Department for Trade and Industry, *New Dimensions for Manufacturing – A UK Strategy for Nanotechnology* (2002) 34.

success are all taken as policy givens. The appropriation of these ideals by nanotechnology policy may well be unintended, but it is an inevitable result of the continued reliance on existing measures. Moreover, the pursuit of these aims is a foregone conclusion and less open to scrutiny as a result.[89] Initiatives to increase investment in nano-enhanced products and production processes have, on the whole, been left unquestioned, except where they have been criticized for committing too little funding to the cause.[90] Action plans to multiply budgets are routinely reproduced[91] without efforts to query the certainties they appear to offer. Not only that, other policy objectives are sidelined and the regulatory environment remains unresponsive, hostile, even, to values besides those directly associated with the EU's commitment to freeing trade. Consequently, the continued application of existing consumer protection necessarily involves the replication not only of its extensive coverage but also of its restrictive practices:

> it is vitally important that attention be given to nanotechnology ... What is absent, above all, is the willingness to consider concerns other than safety risks, not least the issue of whether or not new technologies are desirable, or issues to do with people's convictions about life in general. The benefits and possible adverse effects must first of all be considered, in order to prevent choices being made solely on the basis of economic value while the technology is still at an early stage in its development.[92]

Signs that a different tack may be needed are beginning to show, manifesting in calls for more constructive dialogue with consumers[93] on the necessity for, or desirability of, certain applications of nanotechnology[94] and their cumulative impact on society at large. It may transpire, therefore, that resorting to the coverage of existing consumer protection will not in itself be enough to ensure satisfactory regulatory solutions. This observation does not apply only to nanotechnology for there are many other examples of mismatching problems and regulatory responses.[95] It is particularly stark in

---

89 For similar findings with regard to GMOs, see M. Lee, *EU Regulation of GMOs: Law and Decision Making for a New Technology* (2009) ch. 3.
90 Council for Science and Technology, *Nanosciences and Nanotechnologies: A Review of Government's Progress on its Policy Commitments* (2007).
91 For example, Commission, op. cit., n. 84; Commission, op. cit., n. 86.
92 EP Debate, op. cit., n. 16, MEP Bastiaan Belder.
93 European Economic and Social Committee, Opinion on the Communication from Commission on Regulatory Aspects of Nanomaterials [2009] OJ C218/21, para. 2.12.
94 European Group on Ethics in Science and New Technologies, Opinion on the Ethical Aspects of Nanomedicine, Opinion No. 21 (2007).
95 Lloyd's, *Synthetic Biology: Influencing Development* (2009) 4; Lee, op. cit., n. 89, especially pp. 83–7 where Lee discusses the narrowness of policy responses to GMOs which attach greater weight to scientific and technical information than to 'other legitimate factors' relevant to authorization decisions under Regulation (EC) No. 1829/2003 on GM Food and Feed [2003] OJ L268/1, Article 7. The EP has also been critical of the almost exclusive focus on the scientific assessment of risks and

this instance, however, where a new area displaying different types of risks and uncertainties is grafted onto existing legislative frameworks. Taking into account the range of potential difficulties encountered, the idea of 'existing frameworks' begins to look complex. It may be that prior regulatory provisions are not enough, or it may be that the underlying strategy is questionable or even inappropriate, or it may be both. That the regulation of nano-technologies has certain regulatory and policy predispositions is hardly surprising, given that nanotechnology is currently only being used to produce variations in familiar, regulated products. Yet the enduring effects of prior approaches can be seen in a further sense too, beyond the practice of having recourse to existing legislation. Even where regulations have been amended to account for the novel characteristics of nano-materials, preceding regulatory frameworks and contexts continue to have significant influence. As the following section shows, the internal market project still provides a powerful pull on newly adopted provisions.

## III. NEW MEASURES, OLD WAYS

So far it has been argued that the regulation of nano-products is a job performed by former legislation without any real attempt to take cognisance of the suitability of both the rules and the traditions that are inherent in such an approach. Although the presumptive deference to existing regulatory measures continues to be the approach most frequently applied in the regulation of nanotechnologies, nano-specific provisions are beginning to be adopted in some commercial sectors.[96] For reasons explored below, most debate has centred on mandatory nano-labelling. This is perhaps surprising given that, of all the possibilities, nano-labelling is arguably one of the most obvious and least objectionable means of nano-specific legislation, especially given EU policy on improving consumer information. In many respects, nano-labelling offers a convenient compromise between calls for the regulatory status quo on the one hand, and calls for a differentiated approach to nano-materials on the other. Though it may have other benefits, one of the advantages of nano-labelling is that it is easily integrated into existing regulatory frameworks. Rather than being introduced as a stand-alone requirement, nano-labelling is inserted into existing measures as a new clause. This feature is important because nano-labelling will sustain the aims and functions of its policy predecessors. In this regard it is not so much

---

the promotion of internal market over other socio-economic and environmental objectives, see EP, Resolution on the proposal for a Regulation as Regards the Possibility for the Member States to Restrict or Prohibit the Cultivation of GMOs in Their Territory, P7_TA(2011)0314.
96 The first of these to be introduced was the Food Additives Regulation (EC) No. 1333/2008 [2008] OJ L354/16, Article 12.

responsible for the construction of a new legal reality as it is for perpetuating old, established ones.

### 1. The label 'nano'

Until now, consumers have faced the problem that nano-material ingredients, given their extremely small size, are impossible to detect unless manufacturers voluntarily disclose their presence. Persistent uncertainties about the potential effects of nano-materials together with disagreement over the very definition of 'nano' has meant that disclosure initiatives to date have been sporadic, uncoordinated, and largely peripheral to the core business of regulating possible harms. Whether or not such disclosures ought to be made as a matter of course has been subject to lively debate in recent years, although there is now a growing consensus amongst policy-makers, in the EU and in other jurisdictions besides, that consumers ought to be told whether the products they buy comprise materials in nano-form.

Soon, a range of consumer products containing nano-materials will be required by law to carry the label 'nano' before being introduced to the EU market. EU cosmetics legislation, which in its former guise made no reference to nanotechnology, has already been amended to include the requirement that 'All ingredients present in the form of nanomaterials shall be clearly indicated in the list of ingredients. The names of such ingredients shall be followed by the word 'nano' in brackets'.[97] Similar provisions, even if their precise wording differs, have been proposed in the review of legislation on foodstuffs (generally and in relation to novel foods),[98] biocidal products,[99] and electrical and electronic equipment.[100]

Their introduction has not found favour with all stakeholders. Objections have been raised on several grounds, but most criticisms centre on the pseudo-scientific descriptor 'nano' and its limited capacity to communicate anything meaningful. During negotiations on the insertion of the nano-labelling requirement into the recast Cosmetics Regulation, the Council noted that '[s]ome delegations express doubts on added value of this

---

97 Cosmetic Products Regulations, op. cit., n. 24, Article 19(1)(g).
98 Commission, Communication on the Functioning of the European Union concerning the position of the Council at first reading on the adoption of a Regulation on the provision of food information to consumers COM(2011) 77 final, 5; EP, Resolution on Novel Foods, P7_TA-PROV(2010)0266, Amendment 75 Article 9(2)(c). Note that conciliation talks between the EP and the Council failed, hence the Novel Foods Regulation (EC) No 258/97 remains in force.
99 EP, Report on the proposal for a regulation concerning the placing on the market and use of biocidal products, A7-0239/2010, Amendments 183 and 213.
100 EP, Report on the proposal for a Directive on the Restriction of the Use of Certain Hazardous Substances in Electrical and Electronic Equipment (Recast), A7-0196/2010, Amendment 15.

provision'.[101] The concerns is that the systematic labelling of nano-material ingredients is of little practical use whilst definitions of 'nano-material' remain uncertain[102] and unless the label 'nano' is accompanied by some indication that the product may behave differently as a result. Some have suggested that nano-materials ought to be labelled only when they 'change the characteristics of the product',[103] whilst others believe that the label 'nano' should include additional information 'to indicate in the list of ingredients that such ingredients are present is acceptable to the Commission'.[104] The concern otherwise is that the nano-label will be interpreted as a warning, even if there is no elevated risk involved.[105]

Arguments in support of nano-labelling have gained more traction in the policy debate, in part, no doubt, because product labelling has a long tradition in EU consumer protection. Indeed, the rights to know and to choose are central to EU consumer policy.[106] The overriding aim is to ensure that consumers are free to make purchasing decisions 'in full knowledge of the facts'.[107] As a result of a sustained programme of disclosure measures, European consumers can expect to find a wealth of information about various characteristics of a product displayed on its packaging, such as its origin, quality, safety, intended use, environmental impact, storage, expiry, method of production, composition, and content. This opens up the possibility for consumer choice but also a more robust notion of consumer voice through the

---

101 Council, 'Working Document from General Secretariat of Council/Presidency to the Working Party on Technical Harmonisation on the Proposal for a Regulation on Cosmetic Products (Recast)', 6972/09 LIMITE, 74.
102 For recent attempts to improve clarity, see Scientific Committee on Emerging and Newly Identified Health Risks (SCENIHR), 'Scientific Basis for the Definition of the Term "Nanomaterial"' (2010); and Commission, 'Recommendation on the definition of Nanomaterial' 2011/696/EU.
103 Council, op. cit., n. 101.
104 Commission, op. cit., n. 98.
105 The increasing use of 'nano-free' labels is indicative of concerns that nano-ingredients may cause anxiety among consumers, see Science and Technology Committee, *Nanotechnologies and Food*, HL (2009–10) 22-II, 139. Similar debates have been had in other domains too, notably the policy debate on genetic modification and the EU obligation to label all food and feed products consisting of or containing genetically modified organisms, Regulation (EC) No. 1830/2003 concerning the traceability and labelling of genetically modified organisms and the traceability of food and feed products produced from genetically modified organisms [2003] OJ L268/24, Regulation 6. For commentary, see B. Roe and M.F. Teisl, 'Genetically Modified Food Labelling: The Impacts of Message and Messenger on Consumer Perceptions of Labels and Products' (2007) 32 *Food Policy* 49–66.
106 Treaty on the Functioning of the European Union [2008] OJ C115/47, Title XV, Article 169; see also EP, Resolution on the Commission Communication on 'Consumer Policy Strategy 2002–2006', P5_TA(2003) 0100, paras. 25–6 and 29.
107 E. Stokes, 'You Are What You Eat: Market Citizens and the Right to Know About Nano Foods' (2011) 2 *J. of Human Rights and the Environment* 178–200, at 190.

mobilization of civic and political interests,[108] because '[p]utting consumers in the driving seat benefits citizens'.[109] Nano-labelling is similarly justified, on grounds that consumers 'might like to know whether a food has been produced by the use of nanotechnologies'[110] and, further still, that they have a right to know that a specific product contains nano-materials.

The introduction of nano-labelling measures may not have been altogether straightforward but there is a certain inevitability about it which can be attributed to the hospitable regulatory landscape to which it was introduced. Since existing regulations are well accustomed to the methods and demands of product labelling, and because labelling and information disclosure are consistent with EU consumer policy and tend to be portrayed as benign forms of regulation,[111] nano-labelling presented policymakers with what was arguably the line of least resistance. Yet, in inheriting a fertile soil in which proposals were bound to thrive, nano-labelling has also become subject to the ideological narrowness of the existing regime. It inherits an entrenched bias towards the single market imperative that underpins Community consumer policy. Nano-labelling has so far been adopted or proposed in five product sectors. In each instance, the proposals seek to update the legislation by including, among other things, a provision on mandatory nano-labelling. In each instance, the legislative revisions have been undertaken[112] or proposed[113] using Article 114 TFEU (internal market) as their legal basis. Nano-labelling, therefore, is being introduced to regulatory environments whose market contours are already mapped out.

This gives some flavour of the single market imperative that underlies nano-labelling, although the prevalence of Article 114 TFEU is not on its own a sign that nano-labelling will serve to promote free trade over other policy goals. In fact, the rationale behind nano-labelling, as it is presented in pre-legislative negotiations, may be construed in rather different terms, as an

---

108 See, generally, M. Micheletti, *Political Virtue and Shopping: Individuals, Consumerism, and Collective Action* (2003).
109 Commission, op. cit., n. 45, p. 13.
110 EP, Draft Legislative Resolution on the Proposal for a Regulation on Novel Foods, A6-0512/2008, Amendment 59.
111 Yet information provision can be anything but straightforward and benign. For insights into the complexities of transparency, see E. Fisher, 'Transparency and Administrative Law: A Critical Evaluation' (2010) 63 *Current Legal Problems* 272–314; S. Jasanoff, 'Transparency in Public Science: Purposes, Reasons, Limits' (2006) 69 *Law and Contemporary Problems* 21–45.
112 Cosmetic Products Regulation, op. cit., n. 24, Preamble.
113 Commission, Proposal for a Directive on the Restriction of the Use of Certain Hazardous Substances in Electrical and Electronic Equipment COM(2008) 809 final, 7; Commission, Proposal for a Regulation on Novel Foods COM(2007) 872 final, 5, and Commission, Proposal for a Regulation on the Provision of Food Information to Consumers COM(2008) 40 final, 9; Commission, Proposal for a Regulation Concerning the Placing on the Market and Use of Biocidal Products COM(2009) 267 final, 10.

attempt to inject 'utmost caution'[114] and to provide consumers with reasons not to purchase nano-products by allowing them to 'choose whether or not to take the risk'.[115] But even if nano-labelling was intended as a gesture towards the broader implications of nanotechnologies, its free market subtext has remained substantially intact. Without orchestrated engagement activities or the provision of further information, the label 'nano' does little to enable consumers meaningfully to differentiate between nano and non-nano products, whether on the grounds of risk, uncertainty or any other social or ethical repercussions. What it does do, if anything, is bring nano-products into mainstream commerce. It 'normalizes' the marketing of nano-products by locating them in the same market, under essentially the same conditions, as conventional products. Unless the nano-label is given additional work to do, to re-channel consumer choice through routes other than market transactions, to address the deficiencies in the informational and choice settings, and to counteract the market's status as the ideological driver, it will do little more than reinforce the aims and functions of product labelling generally. These functions are many and include improved flows of information, transparency, and free choice, and they offer a double dividend: the elimination of variations (between member states, between EU citizens, and between consumers) that threaten to undermine single-market opportunities.

These practical and ideological traditions undoubtedly limit the degree to which regulations – even those that are tailor-made for nano-materials – can be implemented on a case-by-case basis, taking into account the particular characteristics of individual nano-products and the needs of the nano-consumer. They can also be said to discourage 'really responsive regulation'[116] in that they are currently programmed to react only to issues thrown up by conventional products. The introduction of nano-specific measures has doubtless prompted important debate on the shape of the regulatory landscape; however, the problem remains that surface-level amendments may not be enough. A deeper excavation is needed to reveal if and how the implementation of those amendments is constrained by the policy substructure and whether the regulation is 'bounded' as a result.


## CONCLUSION

This article has shown how nanotechnologies inherit the regulatory environment in which they now find themselves. The application of existing legislation to nanotechnologies plays out in two different ways. First, it

---

114 EP Debate No. 9 of 7 July 2010, MEP Elisabetta Gardini.
115 id., MEP Sari Essayah.
116 R. Baldwin and J. Black, 'Really Responsive Regulation' (2008) 71 *Modern Law Rev.* 59–94.

entails the application of specific rules and standards designed to deal with conventional types of products. This virtually automatic process may be problematic where nano-scale counterparts present new, sufficiently different, risks and concerns. Where it is recognized as a problem, it is presented in the policy discourse as something that can be overcome by new, tailor-made legislation.

The second way in which existing legislation applies raises a more intractable issue, where former policy substructures and assumptions continue to dominate understandings of how best to respond to nanotechnology. These aspects are more deeply embedded in the legislative regime and more resistant to change as a result. One such example is the promotion by existing measures of the internal market project. Consumer protection legislation in the EU has always had in its sights the development of open and competitive markets, and, because of this, it was always going to be difficult to question the commercial exploitation of nano-products. This has been reinforced by the EU's continuing efforts to secure prominence in the global nanotechnology market. Against this backdrop, debate about the desirability of nanotechnologies and nano-products has largely been bypassed. The commercialization of nano-products is presented in the policy debate as a cut-and-dried eventuality with the upshot that opportunities for reflecting on the social or ethical ramifications of such innovation have been sidelined.

Recently, EU institutions have responded by introducing new, nano-specific provisions into legislation covering a range of consumer products. For instance, some nano-products will be required to carry the label 'nano' as a condition of market entry. Yet, as it is shown here, the simple insertion of legislative text cannot compete with the prevailing culture of markets and consumption. This suggests that the consequences of applying regulatory predecessors may extend beyond a mismatch between old rules and new risks, resulting in more deeply-seated issues of ill-fitting regulatory orientations and foreclosed opportunities for deliberation. New technologies will not always warrant new responses, but their emergence does provide a welcome opportunity to think hard about the tendency, tempting though it may be, to place quite so much faith in a new technology's regulatory heritage.

# The Emergence of Biobanks in the Legal Landscape: Towards a New Model of Governance

EMMANUELLE RIAL-SEBBAG* AND ANNE CAMBON-THOMSEN*

*Biobanks are increasingly seen as new tools for medical research. Their main purpose is to collect, store, and distribute human body materials. These activities are regulated by legal instruments which are heterogeneous in source (national and international), and in form (binding and non-binding). We analyse these to underline the need for a new model of governance for modern biobanks. The protection initially ensured by respect for fundamental rights will need to focus on more interactions with society in order to ensure biobanks' sustainability. International regulation is more oriented on ethical principles and traces the limits of the uses of genetics, while European regulation is more concerned with the protection of fundamental rights and the elaboration of standards for biobanks' quality assurance. But is this protection adequate and sufficient? Do we need to move from the biomedical research analogy to new forms of legal protection, and governance systems which involve citizens?*

## INTRODUCTION

Among the many roles and responsibilities attached to biobanking in biomedical research, their governance is of paramount importance. As biobanking developed initially in the context of clinical trials or as a by-product of clinical activities, its normative and practical framework has evolved

---

\* UMR U 1027, Inserm, Université de Toulouse 3 – Paul Sabatier, Epidémiologie et analyses en santé publique: risques, maladies chroniques et handicap, Département d'épidémiologie et de santé publique, Faculté de médecine, 37 allées Jules Guesde, 31073 Toulouse, France
emmanuelle.rial@univ.tlse3.fr    anne.cambon-thomsen@univ.tlse3.fr

We gratefully acknowledge funding from the European Community's Seventh Framework project, Biobanking and Biomolecular Resources Research (BBMRI), grant no. 212111, and Biobank Standardization and Harmonization for Research Excellence in the European Union (BioSHaRE-EU), grant no. 261433.

logically through the derivation and adaptation of existing biomedical regulatory frameworks applying to research on people. However, the explosive development of biobanking in the last ten years questions this framework; new principles and tools need to be generated. Biological samples are not persons but human materials; biological samples without data attached to them are not very useful for research; the research time frame for projects using biobanks is much longer than that of any individual project, and multiple uses of biological samples over years has become the expected destiny of biobank samples. This phenomenon has been analysed by various disciplines: sociology,[1] philosophy,[2] philosophy of law, political science,[3] economics,[4] and ethics.[5] So, biobanks have become the object of much attention, most certainly with regard to the legal and ethical stakes raised by the use of human samples[6] and of their associated data.[7] In this article we will show why a new model for governing biobanks is needed, using elements from different perspectives. In particular, we will insist on the need for the elaboration of a new model of governance, to be supported by agreed ethical principles and strong public participation. Many questions arise regarding the relevance, or the limitation of existing frameworks that have hitherto served as references for biobank regulation, based partly on analogies with other kinds of biomedical research. We will describe the context of biobanking research as part of the biotechnology landscape, the background and origin of the most commonly used regulations, and discuss new challenges faced when biobanks develop at the level of supra-national research infrastructures. We will conclude by indicating possible directions for the future.

## THE LOCATION OF BIOBANKS AMONG BIOTECHNOLOGIES

Our initial assumption, in light of emerging practices in the field of biotechnology and especially in biobanks, is that, on the one hand, the rule of law is perhaps not the most efficient tool for regulating biotechnology. Other

---

1 V. Tournay (ed.), *La gouvernance des innovations médicales* (2007).
2 B. Hofmann, J.H. Solbakk, and S. Holm, 'Mapping the Language of Research Biobanking: An Analogical Approach' in *The Ethics of Research Biobanking*, eds. J.H. Solbakk, S. Holm, and B. Hofmann (2009) 145–58.
3 H. Gottweis and A. Petersen (eds.), *Biobanks: Governance in comparative perspective* (2008).
4 I. Hirtzlin, N. Preaubert, and A. Charru, 'Analyse de l'activité et du coût des collections de matériel biologique' (1999) 17 *J. d'Economie Médicale* 3–11.
5 A. Cambon-Thomsen, E. Rial-Sebbag, and B.M. Knoppers, 'Trends in ethical and legal frameworks for the use of human biobanks' (2007) 30 *European Respiratory J.* 373–82.
6 Most of the time these elements are legally seen as the entire human person.
7 These data raised questions related to protection of private life and confidentiality.

standards, possibly more flexible, could provide such regulation (ethics, guidelines, professional recommendations, and so on). On the other hand, the inadequacy of traditional regulatory frameworks has resulted in a reluctance among lawmakers regarding the definition of these objects and structures as well as their status, thus creating ambiguity and legal instability in all aspects. Thus, we first have to demonstrate how biobanks are recognized as a tool for producing scientific knowledge, then identify biobanks as a new instrument of biotechnology, and, finally, clarify the rights of 'source-persons'.

*1. Biobanks: a tool for producing scientific knowledge*

The collection of living organisms and tissues has been carried out for many years. Initially centred on their educational value, collections of human biological samples have increasingly become part of medical research protocols. This activity was started, and then structured, primarily in the hope of developing new diagnostic or therapeutic strategies, or of identifying new genes involved in the development of disorders, eventually leading to the possibility of medical intervention.

Although initially centred on the collection of data and eventually of blood samples, research activity has been transformed by the discovery of DNA and by the development of techniques enabling better harvesting, preservation, and use of such biological material. The eventual aim, through the use of biobanks, is either to produce scientific information (in particular genetic) which can be used without any immediate medical application (for example, in population genetics), or to generate information useful for the development of biological tests (including genetic tests) which can be used in medicine, or will give rise to the understanding of disease mechanisms which would indirectly lead to medical intervention.

The growth in the use of collections of human biological samples in research is the fruit of a dual development,[8] both quantitative and qualitative. At the quantitative level, there are more and more samples and, due to the new technologies developed, notably in genetics, these generate more and more information.[9] Thus, progress born of data processing and techniques of cryobiology and molecular biology enables the preservation of a growing number of samples and data, and also an improvement in the duration of this preservation. From this better preservation of biological samples and associated data, the question arises of access to, and sharing of, these collections by the relevant scientific community. At the qualitative level, technical progress means that the information associated with biological samples is more detailed and can be used more optimally. Its use in the health field is

---

8 A. Cambon-Thomsen et al., 'Biobanks for genomics and genomics for biobanks' (2003) 4 *Comparative and Functional Genomics* 628–34.
9 This massive production of data attached to human biological samples calls into question their use for health purposes or for research.

evident; these data are not only used in research, but also enable (notably due to the use of biomarkers) the 'definition of a disease, to pick out the early symptoms, to assess exposure to environmental or nutritional factors, to look at factors of susceptibility, in particular genetic.'[10]

At the same time as technical progress was made with samples, scientists were confronted with diverse banking practices often considered to be a curb on their activities, particularly regarding collaboration. This, then, posed questions of organization and of standardization of sample banks[11] as well as access, for which the public authorities have been mobilized in order to provide solutions. So, the emergence of biobanks as innovative objects engaged all the actors in health research before becoming 'hard notions' in existing laws.

## 2. Biobanks as an element of biotechnology

Biotechnologies as such are not the object of law. They involve 'the application of science and technology to living organisms as well as parts, products and models thereof, to alter living or non-living materials for the production of knowledge, goods and services.'[12] More recently, the European Commission has issued the following more precise definition of what is covered by the term 'modern biotechnology': 'Modern biotechnology can be defined as use of cellular, molecular and genetic processes in production of goods and services.'[13] These applications are, by nature, multiple. Biobanks can be defined as infrastructures, the aim of which is to store, organize, use, and make available biological samples and their associated data for scientific research.[14] These samples must be obtained whilst respecting the rights of the persons from whom they are harvested,[15] termed 'source-persons',[16] and must meet technical requirements in order to be scientifically useful[17]

---

10  INSERM, 'Les collections de ressources biologiques humaines' *Collection Repères*, February 2005, 4.
11  id.
12  OECD, *Glossary of Statistical Terms* (2001), at <http://stats.oecd.org/glossary/detail.asp?ID=219>.
13  E. Zika et al., *Consequences, Opportunities and Challenges of Modern Biotechnology for Europe* European Commission, JRC and IPTS Reference Report EUR 22728 EN (2007).
14  This study focuses on research biobanking only. For the purpose of this article, we exclude therapeutic biobanks and forensic biobanks. Regarding the latter, see A.M. Duguet, E. Rial-Sebbag, M.C. Lacore, and A. Cambon-Thomsen, 'Forensic DNA analysis and biobanking in France' in *New Challenges for Biobanks: Ethics, Law and Governance*, eds. K. Dierickx and P. Borry (2009).
15  Notably, after giving them clear information and after they express their consent.
16  For definition, see p. 118 below.
17  OECD, *Guidelines on Human Biobanks and Genetic Research Databases* (2009), at <http://www.oecd.org/dataoecd/41/47/44054609.pdf>. France has recently adopted a standard governing the quality of Biological Resource Centres: see *Qualité des centres de ressources biologiques (CRB) – Système de management d'un CRB et qualité des ressources biologiques d'origine humaine et microbienne* (2008) NF S96-900.

(quality of samples, biological safety, quality of preservation, traceability, quality of associated annotations and information, that is, computerized database security). Samples can then be gathered into a collection and stored in structures with a high degree of organization: biobanks. The quality of the internal organization of the biobank,[18] as well as its ability to exchange samples, will determine its success, its credibility, and its development. The appearance of these new activities on the biotechnology scene has moreover given rise to new requirements, notably the necessity to fulfil all the relevant administrative obligations. These aspects of quality, although essential in the structuring of the biobank, are not sufficient to ensure its legality: they need to be complemented by consideration of various ethical and legal constraints. An initial overview shows that the existing legal framework is fragmented and a certain amount of mental gymnastics is needed to enable identification of the obligations of the various parties concerned. Although progress of scientific knowledge is presumed to be the common denominator in the motivation for being involved in a biobank, those involved have developed varied interests which need to be reconciled. The aim of *researchers* (and their institutes) is to place the biobanks at the service of medical and scientific biological knowledge, but also at the service of patients in order to deliver, or to confirm, a diagnosis, or to adapt a treatment. *Institutional* policies (of research institutes but also of government ministries) increasingly financially support programmes where the aim is to set up a biobank, whether to enable it to meet the quality criteria required for certification, or to encourage its use in research. The biobank is thus not an isolated object since a number of actors contribute to it in various degrees.

These multiple uses cannot be achieved without a *scientific and ethical evaluation* of the projects. Various structures are typically put in place for such evaluation: external structures enable projects to be scientifically anchored with a recognized methodological approach, and an 'ethical' use of the biological samples. Research ethics committees are mainly responsible for this evaluation. In addition to external evaluation committees, internal committees have the same type of functions[19] (either a single committee or two separate committees). These internal committees are in charge of evaluating the biobank project as a whole and the research projects submitted that require its use. These aims are different in nature since, in the first case, it will be a matter of organizing relations between the biobank and the persons recruited to ensure an ethical basis to the construction of the

---

18 This means the identification of the different responsibilities within the structure, the implementation of the scientific and ethical structures to provide samples to external researchers, and so on.
19 For example, see the United Kingdom Biobank website for the aims and standards set by its Ethics and Governance Council, in particular, the rules for ensuring ethical recruitment and the ethical policy for the provision of samples to researchers, at <http://www.ukbiobank.ac.uk/ethics/intro.php>.

project. Ethical concerns here will essentially be related to the recruitment of participants. In the second case, the committees will be scrutinizing the ethics with regard to access of research scientists to samples and data.

3. *The rights of the 'source-persons'*

In both cases considered above, it is on the two pillars of bioethics, consent and a guarantee of confidentiality with regard to stored data, that ethical evaluations will predominantly rely. It is, therefore, respect for the rights of *source-persons* which is at the heart of the establishment of a biobank. The concept of source-persons is emerging from the confusion regarding the legal and ethical requirements attached to the use of samples. The legal qualification of the source-person cannot be the same when samples are obtained within the healthcare context or within the research context. On the one hand, when samples are used for diagnosis or treatment, people are considered as patients and are protected through the implementation of patients' rights. On the other hand, when samples are gathered from research participants, these persons are considered as donors of body elements and are specifically protected by the rules governing donation, which are quite different from patients' rights.

This clear boundary between care and research becomes blurred when using human bio-specimens. These samples are usually used in the context of medical care, but as this 'material' is also precious for research, these samples are requalified to be used in research. People from whom these samples are removed, the source-persons, usually do not perceive their change of status (patients and/or donors) or the rights attached therein. This article therefore is mainly concerned with this somewhat particular biological material because of its semi-permanent link with the human person.[20] Since personal data are attached to the samples, even if they are coded, they are still considered a part of the person. As a result, from a legal point of view, people assume the same personal rights when talking about biological material or talking about data. The source-persons can claim a right of control on the future uses of their biological samples. Various European legislation translates this right, maintaining, somewhat artificially, a strong link between the sample and the person in order to ensure the effectiveness of this right. But these analogies between person, body elements, and personal rights create an ambiguous situation regarding the reality of the link between the person and the bio-specimen. Samples vary in their uses. They circulate, can be coded in many different ways, and can be used by various researchers. For all these reasons, it is not realistic or practical to ensure people have an actual right of control over them.

---

20 We will not focus on data attached to samples, except in the manner they are linked to the source-person.

So, as biobanks are at the intersection of multiple issues – between medicine (care) and science (research), between market (private interest) and public health (public interest) – several questions arise as to what type of governance will offer supervision of these issues. What should be the level of regulatory response? What role should decision makers play? What types of measures should be favoured, incentives or restrictions? How are these responsibilities distributed? All these questions require both retrospective analysis of the legal and ethical framework for governing biobanks and evaluation of the relevance of these existing texts and guidelines to this new method of performing research.

## A RETROSPECTIVE ANALYSIS OF A CONCEPTUAL FRAMEWORK FOR GOVERNING BIOBANKS IN EUROPE

The process of producing various norms in the field of biobanks is complex, but quite different when considering techniques or concepts. Harmonization is often used in technical fields where standardization procedures are an accepted mechanism, and this is applicable to the field of biobanks. On the other hand, in conceptual fields such as ethics, harmonization is questioned,[21] particularly regarding whether it is desirable or possible without distorting the underlying principles on which ethics are based.

Thus, without denying the existence of other rules, such as those of professional conduct,[22] our study focuses on the classic links between ethics and law, in medicine and biomedical research. These rules, where the lines of demarcation are sometimes blurred, maintain these close links which are historically well described in biomedical research. They cover the entire medical activity from care to prevention, leading to various questions: simple (covering a single activity) or complex (broader questioning including several fields). This link between ethics and law is at the centre of many evolving debates around the new place of the medical practitioner in society, in particular, the new relationships linking the medical establishment to the patient, illustrated by the contemporary shift from paternalism to partnership. The reflection on ethics with regard to these new themes needs to be multidisciplinary and to involve public figures from different perspectives in order to be useful in the supervision of scientific progress. Biomedical ethics has a more strict definition, as the ethics applied to biomedical research and technology. The aim of ethics here is to clarify the values at stake for the various involved parties and to propose methods of resolving tensions raised

---

21 R. Chadwick and H. Strange, 'Harmonisation and standardisation in ethics and governance: conceptual and practical changes' in *The Governance of Genetic Information: Who Decides?*, eds. H. Widdows and C. Mullen (2009) 201–13.
22 We note that there is no real code of conduct adopted for researchers, which could explain the large part of this field covered by Ethics.

by the presence of the different stakeholders with a respect for others in decision making.

With regard to the law, this paper utilizes a very succinct definition. In its wider approach, the law can be defined as a set of rules whose *raison d'être* is to supervise human activities. These rules are variable both in source (national and international), and in form (binding and non-binding instruments). This variability is illustrated at the international level but can also be observed within countries and between countries. There are rules which have a gradually increasing binding force: custom can produce the effects of law but is not written down;[23] *soft law* involves rules which are not binding, including the numerous declarations with regard to biomedical ethics prescribed by international organizations; international treaties and conventions lead to legal obligations for signatory states; and, lastly, European Union law which has the strongest applicability in the national law of member states, notably concerning the application of regulations and Directives adopted by the European Union. At the national level, law is defined as the legal rules in force in a state, commonly called domestic law. The procedures for proposal, discussion, and adoption of these rules are generally inscribed in a constitution or state statutes. Biomedical research continues to waver, between various regulatory approaches: ethics, law, national or international level. From these texts and reflections, an analogy is regularly advanced between biomedical research and research on the human body, which suggests a need for further analysis of these analogical links in the biobank context.

## THE LEGAL AND ETHICAL REFERENCES FOR GOVERNING BIOBANKS

When analysed retrospectively, the legal framework regarding the emergence of biobanks was set up within two periods. First, the founding texts for the protection of persons involved in biomedical research were applied by analogy[24] to human biological samples used in research. So, when it became technically and scientifically possible to use previously stored materials from the human body for research purposes, it was to the protection of the research subjects that regulatory bodies turned to provide the justification for the use of these materials and to provide the source-person with rights. The legal analogies between the human being, the samples from the body, and the rights of human beings have formed the basis for the level of protection

---

23 In life sciences, see C. Byk 'Les sciences de la vie en droit international: de la préhistoire du droit à l'établissement d'une coutume en droit international' (2000) *Gazette du Palais* 1761.
24 E. Rynning, 'Legal challenges and strategies in the regulation of research biobanking' in Solbakk et al., n. 2, pp. 277–313.

afforded to subjects. The more the samples from the body are linked to, or identified with the persons themselves,[25] the more far-reaching are the duties of protection asserted by regulatory bodies. This protection is then principally afforded by respect for fundamental rights. However, samples are different, when considered to be totally detached from the body, and do not permit identification of the subject. These materials are then more easily comparable to biological material and therefore fall under the category of health products. Thus, it is on the nature of the biological materials stored, as well as the type of analysis which may be carried out (notably genetic studies), that the legal supervision of biobanks relies.

Secondly, the legal integration of this new object of biotechnology has been shown in a number of strategies and by application of various types of texts. A study of the various levels of supervision of biobanks at the European and national levels shows that several models of legal policy have been adopted to govern biobanks, but none of them supersedes the others. Thus these models are totally dependent on the legal systems in force in each state and the legal instruments available to the legislator. There are two positions taking shape in EU countries: either specific legislation has been adopted (Estonia,[26] Hungary,[27] Sweden,[28] the United Kingdom,[29] and Belgium having adopted a hybrid model[30]), or provisions with regard to biobanks have been integrated into wider legislative provisions (France,[31] Spain[32]). Despite this disparity of legal regimes, a movement towards governance has been taking shape to supervise the activity of biobanks at both international and European levels. International bodies have suggested models of governance based essentially on the use of Declarations focusing on the supervision of biobanks aimed at genetic research. As for European bodies, they have decided on the protection of human rights and on the quality assurance of biobanks for wider use.

---

25 Notably, with regards to the notion of 'identification'. Thus, if biological elements make persons identifiable, they reveal personal information which is part of private life and which has to be protected as such. This argument can be moderated by the emergence of new technologies (such as whole genome sequencing) which could allow, when crossing with other databases, the re-identification of persons. Nevertheless, even if this possibility exists for genetic data, anonymous material can still be used when all personal links are removed.
26 The Estonian Human Genes Research Act RT I 2000, 104, 685.
27 The Hungarian Parliamentary Act No. XXI of 2008 on the protection of human genetic data and the regulation of human genetic studies, research, and biobanks.
28 The Swedish Biobank Act 2002, 297.
29 The United Kingdom Human Tissue Act 2004.
30 Belgium has transposed the Tissue and Cells Directive (2004/23/CE) OJ L102/48, 07.04.2004, extending its field of application to research activities.
31 Bioethics law no. 2004-800, 6 August 2004, revised on 7 July 2011.
32 Law on Biomedical Research 14/2007 3 July 2007, on Biomedical Research.

## 1. *The international regulation of biobanks: ethics, research, and genetics*

Well before the adoption of texts specific to biobanks, international structures (intergovernmental or professional associations) adopted texts of a general nature for the protection of subjects in biomedical research. The interest of these texts is not limited to their historical nature because they were the first to have been applied regarding the use of samples from the body in research, in the absence of specific texts. It should be remembered that, because of the atrocities committed by the Nazis during the Second World War, and with regard to the role played by certain medical practitioners, when the Nuremberg trials took place, ten principles[33] were proclaimed with regard to biomedical research so as to ensure the respect and dignity of participating persons. All of these principles, far from becoming obsolete, would serve as a reference for future texts. This declaration, incorrectly called a 'code', had the merit of formalizing the spirit reigning at the end of the war regarding biomedical research. But, at the legal level, it only had the value of a declaration and was closer to an ethical opinion. It was quite natural that the following texts were also declaratory in nature: either the professional organizations did not have the power to act otherwise,[34] or the intergovernmental organizations followed a course whereby consensus was more easily reached, with a view to acting less timidly afterwards and using the panel of legal instruments at their disposal.

Amongst the texts adopted by professional organizations, the principal one applying to research on human beings, in which can be found specific provisions concerning human samples, is the Helsinki Declaration. Today this is defined as a declaration of ethical principles, in order to supply recommendations to those contributing to the implementation of a research protocol.[35] Its primary role is to guide the conscience of professionals but in no way does it confine them to a pre-established rule. Although it has no binding value, it is recognized as the gold standard in the management of research involving human beings (particularly with regard to international collaborative research). There is therefore no real correlation between the binding nature of a text and its degree of applicability for research scientists. Biobanks are not expressly targeted by the Declaration (the term is not used in the text). The use and reuse of samples in research is pointed out and must be undertaken after seeking an express informed consent or a waiver from a research ethics committee.[36]

---

33 Including total and legal consent from the subject, voluntary participation, previous experimental results on animals, no risks, investigators' previous experiences in research, liberty to bring the experiment to an end; see *Nuremberg Code*, at <http://ohsr.od.nih.gov/guidelines/nuremberg.html>.
34 These are associations, and thus do not have the capability to conclude treaties
35 The 1964 Helsinki Declaration, amended most recently in Seoul in 2008, at <http://www.wma.net/en/30publications/10policies/b3/index.html>.
36 id., Article 25.

This acknowledges the fundamental link existing between the degree of personal identification of biological materials and the degree of protection afforded to them, as well as the need for an ethical assessment by an independent committee. It should be noted that this text is applicable to all biological samples of whatever nature and that its use in genetics is not specified, unlike several texts adopted by intergovernmental organizations.

Amongst intergovernmental organizations, the principal texts supervising biobanks were adopted, on the one hand, by the United Nations Organization for Education, Science and Culture (UNESCO), and on the other, by the Organization for Economic Cooperation and Development (OECD).

The texts adopted by UNESCO[37] were only declaratory and particularly focused on the need to protect data derived from the genome. First of all, UNESCO insisted on the nature of the genome being part of the common heritage of mankind and, therefore, on the necessary protection of the human species (the Universal Declaration on the Human Genome and Human Rights, adopted 11 November 1997). Secondly, the use of knowledge with regard to the genome was the object of guidelines (International Declaration on Human Genetic Data, adopted 16 October 2003). These texts treat genetic data as separate amongst biological data and consider that they require specific protection.

More focused on questions of fostering uses of biological resources, the OECD centred its proposals on availability of tools common to the Biological Resource Centres (understood in the widest sense, that is, all types of resources: micro-organisms, human, animal or plant cells). These guidelines, adopted in 2007,[38] dedicate a specific chapter to human biological resources and to the scientific use of Biological Resource Centres.[39] The use of resources and data for genetic research is envisaged more strictly in another set of guidelines published in October 2009, for Human Biobanks and Genetic Research Databases (HBGRDs).[40]

Two positions thus appear clearly in the policies of governance proposed by these international structures: a wide option, covering all types of biobanks, for varied use in research; a restrictive option where biobanks for genetic research alone are targeted and which, we believe reinforces the notion of 'genetic exceptionalism'.[41]

---

37 All these texts are available at <http://www.unesco.org/new/en/social-and-human-sciences/themes/bioethics/>.
38 Available at <http://www.oecd.org/dataoecd/7/11/38777441.pdf>.
39 Definition (in French only) at p. 94 of the report: collections are 'Regroupement à des fins de recherche de matériels biologiques sélectionnés sur la base de critères biologiques ou cliniques'.
40 Available at <http://www.oecd.org/sti/biotechnology/hbgrd>.
41 'Genetic exceptionalism' is the concept that genetic material and attached data should be considered apart from the other biological material and data, and should require a higher level of protection because of their nature. For example, see Z. Lazzarini

These texts are not legally binding, but they are extremely important either to affirm common principles (UNESCO) or to standardize practices (OECD). They serve as a reference for scientists and regulatory bodies, and although their violation cannot, in any circumstances, lead to direct sanctions, they can influence the adoption of more binding texts, which is precisely the case with certain texts in European law.

2. *European regulation of biobanks: protection of fundamental rights and quality assurance*

European institutions have appreciated the importance of the health field as a question common to the various member states for a long time. Although all European legal instruments continue to assert strongly the full and complete authority of individual states in the definition and management of their health policies, these same states have widely agreed that in certain specified cases there is strength in unity. This is the case with the instruments developed by the Council of Europe and in community law at the level of the European Union, as described below.

The Council of Europe, an intergovernmental organization for the promotion and the protection of human rights, prepared and adopted the first (and currently the only) contractual instrument with regard to bioethics in Europe. Prepared by the Steering Committee on Bioethics (CDBI), the 'Convention for the protection of Human Rights and dignity of the human being with regard to the application of biology and medicine', adopted on 4 April 1997,[42] is a common base for guaranteeing the rights of human subjects in the face of scientific progress. It took the form of a Framework Convention stipulating the general principles that the signatory parties undertook to implement upon ratification. This Convention is an international treaty having the effect of law within signatory states, and includes additional Protocols bearing on specific subjects.[43] The text came into force on 1 December 1999. The Council of Europe is also the first European intergovernmental organization to propose a common text with regard to collections of human biological samples.[44] Although this text is only a Recommendation, it is integrated into the system of the Convention and its Protocol concerning biomedical research, and currently serves as a reference,

---

'What Lessons Can We Learn From the Exceptionalism Debate (Finally)?' (2001) 29 *J. of Law, Medicine and Ethics* 149–51.

42 Council of Europe, *Oviedo Convention* (1997), STE 164, at <http://conventions.coe.int/Treaty/fr/Treaties/Html/164.htm>.

43 Additional protocols adopted to date are: Council of Europe, Protocol on cloning (1998), Protocol on transplantation (2002), Protocol on Biomedical Research (2005), Protocol on Genetic Testing for Health Purposes (2008).

44 Council of Europe, *Recommendation of the Committee of Ministers to member states on research on biological materials of human origin* (2006), at <https://wcd.coe.int/ViewDoc.jsp?id=977859>, to be revised in 2012.

notably for the protection of the rights of persons in the case of the reuse of material already harvested and preserved. These tools are now references in their field and enable the identification of a common base in bioethics for the various states which are members of the Council of Europe. Despite this strong symbolic and legal value, the regulatory system with regard to bioethics within the Council of Europe is not integrated. Thus, only cases which can demonstrate a violation of the European Convention on Human Rights, which has its own judicial system (the European Court of Human Rights) enable legal proceedings to be brought by citizens seeking justice, and only under certain restrictive conditions.

The situation is different for the texts produced by the European Union. The achievements of the European Union in the field of health are numerous because, although health did not figure as a main heading in the Treaty of Rome, it has been a recognized field of competence since the Treaty of Maastricht.[45] This lack of a formal legal basis has not prevented the European Union (notably with regard to achieving the internal market[46]) from adopting a common health policy, at least in certain fields. There are also specific provisions with regard to research and development which can be found in Articles 163, and following, of the Treaty. Currently, the European Union has a strong presence in the field of public health and research, which as shared fields of competence must meet the principle of subsidiarity (that is, giving precedence to member states). The European Union thus has defined its scope of jurisdiction as well as specific budgets to implement these policies (adoption of Directives). Today these policies provide the Union with common tools for the protection of public health, for example, or for the implementation of a policy of internationally competitive scientific research. These Directives, which are mandatory, enable the obstacles to the free movement of persons,[47] as well as those with regard to the free movement of goods,[48] to be reduced. It is with regard, in particular, to market authorization of medicinal products that the Union adopted the Directive concerning clinical trials,[49] to provide all members with a common basis for clinical research protocols. Although this Directive ensures a high level of protection for research subjects, it focuses heavily on the procedures to be respected before, during, and after a clinical trial.[50] In addition, the

---

45 Art. 152, and now Art. 168.
46 Treaty Article 3 and, more precisely, provisions of the treaty ensuring free movement of persons and goods.
47 As an example, many Directives have been adopted in the field of free movement of medical professionals.
48 Many Directives adopted in the field of drugs
49 Directive 2001/20/EC of 4 April 2001, on the approximation of the laws, regulations, and administrative provisions of the member states relating to implementation of good clinical practice in the conduct of clinical trials on medicinal products for human use, OJ L121/34, 1.05.2001.
50 id.

European Union focuses on the guarantee to patients of access to quality products by proposing the establishment of common procedures ensuring the traceability of products derived from the human body.[51] It should be emphasized that although this legislative arsenal does apply to biobanks storing cells, it only covers those having a therapeutic goal. Therefore, only clinical research where the aim is either to permit individual clinical use (for one patient), or to market a product, is covered; *in vitro* research falls outside the field of the Directive. It is regrettable that this application is restrictive, as it does not settle the questions that arise in the large collaborative research projects which are still searching for common instruments to preserve, exchange, and use biological samples and data. In addition, in discussions of these Directives, the Union has refused a number of times to venture into the field of bioethics, arguing its lack of legal jurisdiction for so doing,[52] which leaves scientists on highly unstable legal ground. However, in addition to this approach to European Community policy, a real reflection should be noted with regard to ethics in sciences as represented essentially by the Opinions given by the European Group on Ethics in Science and New Technologies[53] as well as the resolutions of the European Parliament, the Opinions of ad hoc groups,[54] and the various guidelines and evaluation from the Ethics sector of the Commission for the seventh framework research programme.[55]

---

51 Directive 2004/23/EC on setting standards of quality and safety for the donation, procurement, testing, processing, preservation, storage and distribution of human tissues and cells (2004); Directive 2006/17/EC implementing Directive 2004/23/EC of the European Parliament and of the Council as regards certain technical requirements for the donation, procurement and testing of human tissues and cells (2006); Directive 2006/86/EC implementing Directive 2004/23/EC of the European Parliament and of the Council as regards traceability requirements, notification of serious adverse reactions and events and certain technical requirements for the coding, processing, preservation, storage and distribution of human tissues and cells (2006).
52 Discussions regarding clinical trials and the positions of the European Parliament are available at <http://www.europarl.europa.eu/oeil/file.jsp?id=109002>.
53 Notably Opinion no. 11 – *Ethical aspects of human tissue banking* (1998), available at <http://ec.europa.eu/european_group_ethics/docs/avis11_en.pdf>.
54 E. Mcnally, A. Cambon-Thomsen et al., *25 Recommendations on the ethical, legal and social implications of genetic testing* (2004), at <http://ec.europa.eu/research/conferences/2004/genetic/pdf/recommendations_en.pdf>.
55 G. Chassang, E. Rial-Sebbag, and A. Cambon-Thomsen, 'The foundation of research ethics in community law' (2011) 22 *J. International de Bioéthique* 187–203.

# DISCUSSION: A PERSPECTIVE ON THE ETHICAL AND SOCIAL DIMENSION OF BIOBANKS FOR A GOVERNANCE FRAMEWORK

Whereas regulation is still fragmented at various levels, some common principles towards a new model of governance are emerging from the practices and from the policies proposed, particularly at the European level.

Indeed, while researchers have increasingly used their biological samples in practice,[56] the use of biobanks has also been widely promoted by European policymakers. Thus, in the move to create European infrastructures,[57] specific funds were allocated to organize a network of European biobanks in 2008,[58] suggesting that ethical issues will now be considered at this level. Notably through the creation of an infrastructure, the Biobanking and Biomolecular Resources Research Infrastructure (BBMRI),[59] new ethical issues are raised because of its transnational nature. This change of scale poses questions, in particular, regarding mechanisms of regulation. Indeed, if the traditional questions about the ethics of biobanks[60] continue to be debated, the possible 'harmonization' of European ethics is now opened for debate. Ethics, whether biomedical, medical or research, is characterized by its strong cultural dependence. Now the challenge for these new structures is to enforce these differences whilst promoting common standards. These standards do not respond to the traditionally accepted notion referring to technical standards but to new approaches to create harmonization.

This new governance,[61] at a supranational level, allows a large role for ethical reflection in the absence of collectively applicable legal rules. This discussion began at national level in the various ethics committees which gave opinions on biobanks. However, a comparison of these various opinions[62] shows that the scope of ethical reflection at European level is varied and the elements necessary for an acceptable ethical use of samples vary from state to state. Here, we are certainly facing a new form of ethics in the life sciences area; after what authors called 'applied ethics',[63] a new

---

56 This point is currently not based on scientific data on the level of uses of biocollection in research. That is the reason why we adhere to the proposal of implementing tools to generate such data: see A. Cambon-Thomsen, 'Assessing the impact of biobanks' (2003) 34 *Nature Genetics* 25–26.
57 For the European initiative on infrastructures see <http://cordis.europa.eu/esfri/roadmap.htm>.
58 M. Yuille et al., 'Biobanking for Europe' (2008) 9 *Brief Bioinform* 14–24.
59 BBMRI, at <http://www.bbmri.eu/>.
60 M.G. Hansson, 'Ethics and Biobanks' (2009) 100 *Brit. J. of Cancer* 8–12.
61 M. Mayrhofer and B. Prainsack, 'Being a member of the club: the transnational (self-) governance of networks of biobanks' (2009) 12 *International J. of Risk Assessment and Management* 64–81.
62 BBMRI report on ethics policies relating to biobank in the members of the project, issued in 2010, not yet published.
63 R.F. Chadwick, *Encyclopedia of Applied Ethics* (1997).

'organizational ethics' is emerging. This new form of ethics will embody new features: more flexible than the law and less technical than the standard, it is certainly an intermediate norm to be referred to for constant adaptation to the needs of regulation in the field of biobanks. This new approach emphasizes the limits of legal harmonization because of the technical limitations posed by the legal instruments themselves, and it allows space for normative creativity. The only proposals that can be made for a 'harmonization of ethics', understood as a tool of governance, should be based on these common principles for the protection of participants, taking into account specific issues relating to biobanks (definition, informed consent, transfer samples and data, future uses, and so on) and, more particularly, should take into consideration the involvement of the public.[64]

We have mapped the biobanking field in the context of biotechnologies and described the historical development of the present normative frameworks in the ethical and legal dimensions that apply to biobanks. Before concluding, we would like to open up reflection on the social dimension of biobanks that has been relatively ignored until recently. This represents an essential ground for the development of biobanking activities. What do the people think? What is important from the point of view of participants and how might these social considerations influence biobanking activities or give insights for developing a framework that would embrace the full landscape of stakeholders?

With this aim, a number of studies have been conducted recently in the context of the biobank and biomolecular research infrastructure (BBMRI, FP7) in order to prepare a survey of the public perception of biobanks in Europe.[65] Various strategies to address the question of 'What corresponds to a biobank European infrastructure on the societal level?' and, in the absence of a 'European society', possible strategies to interact with varied 'European publics' were discussed.[66] The basic assumption was that it would be key to interact openly and transparently with the European citizenry. No other strategy would accord with the assembled wisdom of science and society studies on Europe conducted during the last decades. Building on previous work, a group prepared the basis for a European-level survey.[67] Most of the existing data in this domain are for the population of the United Kingdom, the United States, and northern Europe. Thus a more diverse focus had to be worked out and, in addition, almost nothing is currently known about the relationship between biobank projects, their increasingly transnational

---

64 H. Gottweis and K. Zatloukal, 'Biobank governance: trends and perspectives' (2007) 74 *Pathobiology* 206–11.
65 H. Gottweis and G. Lauss, 'Biobank governance in the post-genomic age?' (2010) 7 *Personalized Medicine* 187–95.
66 H. Gottweis, 'Good Biobank Governance: How to Avoid Failure' [*in Chinese*] (2009–10) 30 *Medicine & Philosophy* 8–13.
67 G. Gaskell and H. Gottweis, 'Biobanks need publicity' *Nature*, 10 March 2011, 159–60.

ramifications, and public perception of the transnationalization of biobank research.

This European approach was based on three complementary kinds of methods:

(i) a pilot study in two countries (Austria and the Netherlands) via a focus group approach[68] among different publics. The focus groups (8–12 people in group discussion) explored the topic of biobanks, how biobanks are perceived, how they are evaluated with respect to their purpose, which issues of concern and risk arise, and what benefits are perceived. Based on this pilot study, the two further steps were triggered.

(ii) the proposal of specific questions for a quantitative study in the framework of a Eurobarometer survey, that took place in 2010. Eurobarometer is a series of surveys regularly performed on behalf of the European Commission. It produces reports of public opinion on certain issues relating to the European Union across the member states. The 2010 Eurobarometer survey EB 73.1, 'Life Sciences and Biotechnology' contained 8 questions on biobanks. The results were released in November 2010 and are summarized as follows in the report from the European Commission:

> While approximately one in three Europeans have heard about biobanks before, nearly one in two Europeans say they would definitely or probably participate in one, with Scandinavian countries showing the most enthusiasm. And people do not seem to have particular worries about providing certain types of information to biobanks: blood samples, tissue samples, genetic profile, medical records and lifestyle data elicit similar levels of concern. However, amongst those similar levels there are some nuances. In twelve countries, providing one's medical records provokes the most worry, and in ten countries it is the genetic profile that is most worrying. Asked about who should be responsible for protecting the public interest with regard to biobanks, we find a split between those countries opting for self-regulation (by medical doctors, researchers, public institutions such as universities or hospitals) and those opting for external regulation (ethics committees, national governments, international organisations and national data protection authorities). Broadly speaking, respondents in those countries which show higher levels of support for biobanks tend to favour external regulation more than self-regulation. In those countries where biobanks are unfamiliar, self-regulation is a more popular way of guarding the public interest. On the issue of consent, almost seven in ten Europeans opt for specific permission sought for every new piece of research, one in five for broad consent, and one in sixteen for unrestricted. But of those more likely to participate in the biobank, some four in ten opt for either unrestricted or broad consent.[69]

---

68 M.M. Hennink, *International Focus Group Research. A Handbook for the Health and Social Sciences* (2007).
69 G. Gaskell et al., *Europeans and Biotechnology in 2010: Winds of change?* (2010), at <http://ec.europa.eu/research/science-society/document_library/pdf_06/europeans-biotechnology-in-2010_en.pdf>.

(iii) the setting up of a refined script and methodology of analyses for further focus groups in several countries (Germany, Greece, the United Kingdom, Finland, and France in addition to Austria and the Netherlands). Although this work will be pursued further, the preliminary analyses underline the importance of trust, the preference for narrow consent in general, and the preoccupation frequently encountered regarding privacy and data protection. While people frequently accept the way science develops, there is a general desire for regulation and a focus on governance issues.[70] This is a key point in the evolution of a framework for biobanks.

## CONCLUSION

We have shown that there are limits in using analogies from other contexts of research or health sector activities when regulating biobanks. Taking into account public attitudes and understanding their basis may be a route to construct not only relevant ethical and legal frameworks but, in addition, solid and sound governance systems for the field of biobanking that constitutes a large part of biomedical research nowadays.

---

70 H. Gottweis, H. Chen, and J. Starkbaum, 'Biobanks and the phantom public' (2011) 130 *Human Genetics* 433–40.

# The Legal Landscape for Advanced Therapies: Material and Institutional Implementation of European Union Rules in France and the United Kingdom

AURÉLIE MAHALATCHIMY,*,** EMMANUELLE RIAL-SEBBAG,*
VIRGINIE TOURNAY,** AND ALEX FAULKNER***

*In 2007, the European Union adopted a* lex specialis, *Regulation (EC) No. 1394/2007 on advanced therapy medicinal products (ATMPs), a new legal category of medical product in regenerative medicine. The regulation applies to ATMPs prepared industrially or manufactured by a method involving an industrial process. It also provides a hospital exemption, which means that medicinal products not regulated by EU law do not benefit from a harmonized regime across the European Union but have to respect national laws. This article describes the recent EU laws, and contrasts two national regimes, asking how France and the United Kingdom regulate ATMPs which do and do not fall under the scope of Regulation (EC) No. 1394/2007. What are the different legal categories and their enforceable regimes, and how does the evolution of these highly complex regimes interact with the material world of regenerative medicine and the regulatory bodies and socioeconomic actors participating in it?*

---

\* INSERM, UMRS 1027, et Université de Toulouse, Paul Sabatier – Toulouse III, Faculté de médecine, 37 allées Jules Guesde, F-31073, Toulouse, France
mahalat@cict.fr    rialseb@cict.fr
\*\* PACTE Politique-Organisations, UMR 5194, Institut d'Etudes Politiques/ F-38040 Grenoble, France
virginie.tournay@iep-grenoble.fr
\*\*\* Department of Political Economy, King's College London, Strand, London WC2R 2LS, England
alex.faulkner@kcl.ac.uk

The work presented in this article was supported by THERACELS (The medical uses of human stem cells faced with administrative regulation), ANR Project No. ANR-08-JCJC-0048-01. Faulkner's contribution draws on research funded by the ESRC in awards L218252058 and RES-000-22-1814.

## INTRODUCTION

'Human materials' are increasingly being used in developing medical products under the impetus of the life sciences. Many of these new products are seen as part of the global trend toward 'regenerative medicine', a new paradigm for medicine itself. Such developments attract the attention of law making and regulation, with the goals of protecting and improving public health, ensuring safety, and advancing scientific and industrial ambitions. These developments are producing significant shifts in the relationships between the human and material worlds, bringing them closer together and complicating their distinction. Such shifts have been theorized by sociologists using terms such as 'biomedicalization'.[1] Human tissues, cells, and genes have become the object of regulation worldwide, including new laws in the European Union, which has caused 'a major reshaping of the regulatory landscape of the life-sciences in Member States'.[2] As will be seen in this discussion, the new EU law raises a number of issues that are important to analysis from the perspectives of socio-legal studies and science and technology studies. Conspicuous amongst these is the issue of how legal concepts and regulatory institutions can be 'matched' to the scientific, technological, and industrial categories that emerge in the development and testing of complex new medical materials, and how these vary in the framings of different, bounded legal regimes such as national political cultures, building on existing regimes.[3] Scientists and governments are faced with products derived from living matter, which are, by definition, much more difficult to stabilize than manufactured products using and processing inert materials.

As will be shown below, a new EU 'advanced therapy' regulation defines, more or less, what henceforth does and does not constitute a human materials-derived medical product 'within the law', and it also leaves open some key questions about how certain classes of medical artefact, produced under certain circumstances, might be regulated when they fall *outside* the EU-wide regime that the new regulation established. Crucially, for example, this raises issues of what types of social actors in what types of institutions may participate in the EU regime of regenerative medicine,[4] either as

---

1 A.E. Clarke et al., 'Biomedicalization: Technoscientific transformations of health, illness, and US biomedicine' (2003) 68 *Am. Sociological Rev.* 161–94.
2 M. Favale and A. Plomer, 'Fundamental disjunctions in the EU legal order on human tissue, cells and advanced regenerative therapies' (2009) 16 *Maastricht J. of European and Comparative Law* 89–111.
3 V. Tournay (ed.), *La gouvernance des innovations médicales* (2007).
4 We use the term 'regenerative medicine' to refer broadly to the field under discussion even though it is not equivalent to 'advanced therapy medicinal products'. We do so because it has become a widely used term, especially amongst participating scientists and industry, and is also known to the public, rather than because it has a legal definition. As used here, it may refer to a range of therapeutic applications such as prevention and repair as well as strictly 'regeneration'.

producers or users/consumers. The discussion below will show how the new legal regimes grapple with trying to define delicate distinctions between, for example, what is and is not a 'hospital', what is and is not an 'industrial process' using human materials, and how the borderline is drawn between commodification and non-commodification or commercialization.[5] The new regulation thus enters into the heart of the professional working routines and standards involved in the biotherapeutic manufacturing process. In order to highlight these questions, therefore, we focus in particular in this discussion on the borderline between what is and what is not included in the new EU human tissues and cells and advanced therapy regimes, on some of the most relevant rules regarding clinical trials for new medical entities, and how these matters are being approached institutionally and in terms of the legal conceptualization of human materials, in two contrasting EU countries, namely, France and the United Kingdom.

Perhaps the most conspicuous feature of the legal landscape for the scientific research, materials, and products of regenerative medicine, is its sheer complexity. Much of this complexity is illustrated in the descriptions of relevant laws provided in the body of this paper. To some extent this complexity is ironic, given the aim of the European Commission to provide 'legal clarity' and harmonization within the European area through the development of a new legal regime. The question of the extent to which this complexity and instability is, as it were, the 'natural' reflection of the biologies and ever-evolving living technologies involved, or the extent to which it is more the result of interacting regulatory, social and ethical, and institutional forces, is a key one that the article addresses. Regardless, different countries are tackling the regulation of these materials and associated sciences and technologies in widely different ways, resulting in a segmented marketplace of different regulatory regimes involved in constituting the emerging worlds of the new regenerative medicine paradigm.

The article now turns to consider the recent developments in EU law relating to regenerative medicine in the form of its 'starting materials' of tissues and cells, and the new category of 'advanced therapy' which is the outcome of several years' protracted negotiation between stakeholders in this field.

---

5 In this paper, we consider that the English 'non-commodification' corresponds to the French principle of 'non-patrimonialité' of the human body and its elements, which means that the human body and its elements cannot be the object of a financial agreement.

# REGENERATIVE TECHNOLOGIES AND MEDICAL MATERIALS REGULATED BY EU LAW: LEGAL REGIMES ENFORCEABLE IN FRANCE AND THE UNITED KINGDOM

Human cells and tissues have been identified as a distinct category of medical material for legislation in the EU, and an entirely new category of *product* has also been devised by regulators, namely, 'Advanced Therapy Medicinal Products' (ATMP). This 'advanced therapy' designation does not exist as a legal categorization elsewhere in the world. Advanced Therapy Medicinal Products have been defined in the EU as medicinal products based on genes, cells or tissues. Gene therapy,[6] for example, targeted at cancer cells or aimed to replace defective genes in diseases with genetic causation such as cystic fibrosis, and cell therapy,[7] for example, a patient's cartilage cells extracted, multiplied, and re-implanted, have been regulated as medicinal products under the EU general legal framework since 2003.[8] But tissue-engineered products[9] (TEPs), for example, 'living skin' including a layer of manufactured biomaterial, lay outside any EU legislation.

In Europe, products utilizing human tissues/cells may be regulated either by EU law or by national laws, depending on the determined material and legal character of the 'product' and/or the institutional and work-routine features of the process of preparing it. A comparative approach can shed

---

6 Gene therapy medicinal products means:
   a biological medicinal product which has the following characteristics: it contains an active substance which contains or consists of a recombinant nucleic acid used in or administered to human beings with a view of regulating, repairing, replacing adding or deleting a genetic sequence; its therapeutic, prophylactic or diagnostic effect relates directly to the recombinant nucleic acid sequence it contains, or the product of genetic expression of this sequence. Gene therapy medicinal products shall not include vaccines against infectious diseases.
   Annex, Part IV, 2.1 of Directive 2009/120/EC.
7 A somatic cell therapy medicinal product means a biological medicinal product which:
   contains or consists of cells or tissues that have been subject to substantial manipulation so that biological characteristics, physiological functions or structural properties relevant for the intended clinical use have been altered, or of cells or tissues that are not intended to be used for the same essential function(s) in the recipient and the donor; [and] is presented as having properties for, or is used in or administered to human beings with a view to treating, preventing or diagnosing a disease through the pharmacological, immunological or metabolic action of its cells or tissues.
   Annex, Part IV, 2.2 of Directive 2009/120/EC.
8 Directive 2003/63/EC of 25 June 2003, amending Directive 2001/83/EC of the European Parliament and of the Council on the Community code relating to medicinal products for human use, OJ L159/46, 27.06.2003.
9 A tissue engineered product contains or consists of engineered cells or tissues (of human or animal origin or both, viable or non-viable), and is presented as having properties for human beings to regenerate, repair or replace a human tissue: Regulation (EC) No. 1394/2007 on ATMP, Article 2.1(b).

light on the complexity and flexibility of different national regimes within the over-arching EU risk-based regime.[10] Thus, we will analyse and compare how France and the United Kingdom, both countries with relatively high levels of activity in the regenerative medicine sector, regulate medical materials and advanced therapy medicinal products which do and do not fall under the scope of the new EU laws applicable to regenerative products using human materials.

The structure of the discussion is as follows. First, we explain the legal frame for tissues and cells used in therapy and manufactured as products regulated by EU law, as it is enforceable in every EU member state, noting differences in implementation between France and the United Kingdom. Secondly, analysis will be presented of how each country regulates products which remain *not* covered by EU law. This raises the issue of the 'industrial process' of routine production of therapies, by which the new regulation attempts to distinguish commercial enterprise from 'hospital'-based preparation of therapies for single patients on a one-off basis, a key point of negotiation and conflict during the political discussion of the new products Regulation.

## THE LEGAL FRAME FOR TISSUES AND CELLS USED FOR HUMAN APPLICATIONS AND PREPARED OR MANUFACTURED FOR USE IN OR AS PRODUCTS

Three main fields particularly relevant for companies and other establishments developing regenerative medicine products using tissues and cells have been regulated at EU level: (i) the use of human tissues and cells for human application; (ii) clinical trials; and (iii) marketing authorization and follow-up of regenerative/advanced therapy medicinal products (ATMPs). Below we describe each of these in turn. These sets of laws and regulations are highly complex, and we therefore of necessity summarize the main legal measures and institutional aspects as they refer to the biological materials and products.

### 1. *The EU regulation of tissues and cells as medical materials*

Tissues and cells, when considered as medical materials, have to rely on two pieces of legislation divided according to the chain from procurement to the distribution of the final medicinal products. On the one hand, three EU Directives have been developed to regulate the procurement and use of human tissues and cells for human application. Directive 2004/23/EC of

---

10 See A.-M. Farrell, 'The politics of risk and EU governance of human material' (2009) 16 *Maastricht J. of European and Comparative Law* 41–64.

31 March 2004, often called the "mother directive" (also known in shorthand as the 'Tissues and Cells' Directive, or sometimes, and more informatively, the 'tissue banking' Directive), provides the framework legislation,[11] and two supplementary technical Directives provide detailed requirements.[12] These Directives establish quality and safety standards. On the other hand, where a product enters in the scope of the Regulation on ATMP,[13] the Tissues and Cells Directive will only apply for the donation, procurement, and testing phases of the medical materials. The later steps (processing, storage, and distribution) have to comply with the Regulation on ATMP[14] which concerns medicinal *products* (see below). This means that two complementary sets of rules have to be respected. Interestingly, one main institution, AFSSAPS-French Agency for the Safety of Health Products, will control the enforcement of these rules in France although an opinion may be required from the Biomedecine Agency,[15] whereas, in the United Kingdom, at least three agencies are in place to cover the whole chain of human materials activity: the Human Tissue Authority (HTA) for the regulation of tissues and cells other than gametes and embryos for human application, the Human Fertilisation and Embryology Authority (HFEA)[16] for the regulation of gametes and embryos for human application, and the Medicinal and Healthcare products Regulatory Authority (MHRA) for medicinal *products* based on human body elements.

The directives on tissues and cells have been transposed into British[17] and French[18] laws. In France, AFSSAPS[19] is in charge of the technical imple-

---

11 Directive 2004/23/EC of the European Parliament and of the Council of 31 March 2004 on setting standards of quality and safety for the donation, procurement, testing, processing, preservation, storage and distribution of human tissues and cells, OJ L102/48, 07.04.2004.
12 Directive 2006/17/EC of 8 February 2006 implementing Directive 2004/23/EC as regards certain technical requirements for the donation, procurement and testing of human tissues and cells, OJ L38/40, 09.02.2006, and Directive 2006/86/EC of 24 October 2006, implementing Directive 2004/23/EC as regards traceability requirements, notification of serious adverse reactions and events, and certain technical requirements for the coding, processing, preservation, storage, and distribution of human tissues and cells, OJ L294/32, 25.10.2006.
13 ATMP Regulation, op. cit., n. 9, Preamble (6).
14 id., Article 3.
15 The French Biomedecine Agency was set up by the 2004 Bioethics Law. It is the reference authority for medical, scientific, and ethical aspects, notably those related to procurement and transplant of organs, tissues, and cells: <http://www.agence-biomedecine.fr/>.
16 HFEA, at <http://www.hfea.gov.uk/>.
17 Human Tissue (Quality and Safety for Human Application) Regulations 2007: these Regulations complete the Human Tissue Act 2004 which covers England, Wales, and Northern Ireland (Scotland has separate provision).
18 In France, many legal texts implement those directives: notably, Law n2011-814 of 7 July 2011 on bioethics, French OJ n157, 08.07.2011, 11826, text n1.
19 AFSSAPS, at <http://www.afssaps.fr/>.

mentation of the regulation. French law does not only provide stringent safety requirements for human tissues and cells, it also protects them through fundamental rights and, notably, the principle of respect for human dignity as elements of the human body through the integration of measures protective of the human body within the civil code regarding the respect due to the person.

In the United Kingdom, the HTA[20] detailed in specific guidance the standards required under the Human Tissue Regulations.[21] The overwhelming weight of provision in the United Kingdom's Human Tissue Act is devoted to elucidating the principles of informed consent by patients.

However, the abolition of the HTA and the HFEA is being discussed in the United Kingdom (spring 2011). A previous review in December 2006 proposed to merge the HTA and the HFEA into a single Regulatory Authority for Tissue and Embryos[22] and the positioning of these bodies is again the subject of government attention. This question was debated on 1 February 2011 where a spokesman for the Department of Health stated that his department was 'planning to undertake a public consultation exercise ... about where HFEA and HTA functions are best transferred'.[23] Nevertheless, the existence of two agencies presents the advantage of distinguishing clearly the two sets of rules enforceable for human tissues and cells either as raw materials or as medicinal products once transformed by a biomanufacturing process. But, at the same time, it could be seen as a more complex framework for researchers and manufacturers who might have difficulties understanding the remit of each agency. Equally, in France, it may be considered inconsistent, ambiguous, and potentially a conflict of interest to highlight the significance of the implementation of fundamental rights to human body elements while the same agency also controls both the procurement of human body elements *and* the development of health products based on them. However, the Biomedecine Agency appears to have a safeguard for ethical matters, in particular through its advisory council, even though this is not an ethics committee. Thus, we can see that, whereas the weight of the implementation of the law in France is on non-commodification, in the United Kingdom the major focus is more on the institutionalization of individual consent procedures for allowing but controlling citizens' and patients' rights over use of body materials.

---

20 HTA, at <http://www.hta.gov.uk/>.
21 HTA, *Guide to Quality and Safety Assurance of Human Tissues and Cells for Patients Treatments* (2010); HTA Directions 003/2010 relating to licences granted under the Human Tissue (Quality and Safety for Human Application) Regulations 2007.
22 Department of Health, *Review of the Human Fertilisation and Embryology Act: Proposals for revised legislation (including establishment of the Regulatory Authority for Tissue and Embryos)* (2006; Cm. 6989).
23 Earl Howe, 724 *H.L. Debs* col. GC 343 (1 February 2011).

## 2. EU regulation of clinical trials

Regenerative or advanced therapy products have been deemed to be high-risk products which will therefore require clinical trials to place them in the European marketplace. Directive 2001/20/EC on clinical trials for medicinal products for human use[24] and Directive 2005/28/EC for good clinical practice[25] as regards investigational medicinal products,[26] and authorization of their manufacture or importation, apply to regenerative medicine products. The Regulation on ATMP, detailed below, extended application of the therapy-specific rules to tissue-engineered products[27] and to the adoption of guidelines specific to ATMPs.[28]

Two main specific rules apply to clinical trials involving an advanced therapy product. First, the usual 60-day decision period following a valid clinical trial application can been extended to 90 days for products based on human body elements, and it may be doubled where the consultation of a group or a committee is deemed required by the member state concerned.[29] Secondly, while the non-opposition of the authority is usually sufficient to start a clinical trial, an explicit written authorization is required for trials involving ATMPs or any medicinal product containing genetically modified organisms, indicating the high degree of risk perceived.[30] Apart from EU clinical trials law, France and the United Kingdom each have specific measures regarding the clinical trials of products based on human body elements, which we note below.

In France, EU law on clinical trials applies to medicinal products derived from gene therapy [31] and from cell therapy which are still considered to be medicinal products in French law. Thus it applies to gene therapy, that is, pharmaceutical products and preparations[32] derived from gene therapy, and

---

24  OJ L121/34, 01.05.2001.
25  OJ L91/13, 09.04.2005.
26  Directive 2001/20/EC, Article 2(d), defines an 'investigational medicinal product' as:
    a pharmaceutical form of an active substance or placebo being tested or used as a reference in a clinical trial, including products already with a marketing authorization but used or assembled (formulated or packaged) in a way different from the authorized form, or when used for an unauthorized indication, or when used to gain further information about the authorized form.
27  ATMP Regulation, op. cit., n. 9, Article 4(1).
28  European Commission, Detailed guidelines on good clinical practice specific to advanced therapy medicinal products, 3 December 2009, ENTR/F/2/SF/dn D(2009) 35810.
29  Directive, op. cit., n. 26, Articles 6(7) and 9(4).
30  id., Article 9(6).
31  Article L5121-8 of the French Public Health Code. They are prepared in advance and according to an industrial process. Thus, they are submitted to EU legislation on ATMP and to a centralized marketing authorization.
32  Article L5121-1-12° of the French Public Health Code. They are prepared in advance and to one or several patients on medical prescription. As there is no industrial

to cell therapy,[33] that is, pharmaceutical products derived from human and xenogeneic (animal materials-based) cell therapy[34] and preparations derived from xenogeneic cell therapy.[35] When the research involves genetically modified organisms, the highest possible level of authorization is invoked, so the clinical trial application dossier must comprise the classification of the organism by the Haut Conseil des Biotechnologies (High Council of Biotechnologies – HCB)[36] and the consent of the Ministry in charge of research.[37] However, importantly, preparations derived from human cell therapy are never considered as medicinal *products* in French law. Consequently, they fall under the legal regulations governing human cells and tissues[38] discussed above, including for clinical trials,[39] as distinct from the ATMP *product* Regulation discussed below. The regime for tissue engineering is identical as for cell therapy.

In the United Kingdom, a specific national 'Gene Therapy Advisory Committee' (GTAC) has responsibility for the ethical review of research study involving not only gene therapy, but also embryonic stem-cell therapy, cell therapies derived from stem-cell lines, and the therapeutic use of genetically modified stem cells or therapeutic xenotransplanation. Interestingly, from 1 May 2008, the GTAC can transfer applications to other (locally-based) Research Ethics Committees if a gene therapy proposal is deemed to be 'low genetic risk'.[40] The notion of 'low genetic risk' can be regarded as a technical tool, set up by the GTAC itself. In spite of a 'decision tree'[41] indicating the regulatory routing of products, this notion is not clearly defined. In France, such a concept does not exist, but neither is the range of applications covered by the HCB clearly defined. Although the HCB to date has given opinions on gene therapy only, its mission also covers 'other biotechnologies'.[42] Thus, in legal terms, it seems there is a non-defined

---

process, they are not submitted to the EU legislation on ATMP. The marketing authorization is delivered by AFSSAPS for a specific therapeutic use.
33 Article L1243-1 of the French Public Health Code.
34 Art. L5121-8, op. cit., n. 31.
35 Article L5121-1-13° of the French Public Health Code.
36 The HCB replaces the Commission de Génie Génétique (Genetic Engineering Committee) and the Commission d'étude de la dissémination des produits issus de génie biomoléculaire (Committee for studying the dissemination of products derived from biomolecular engineering).
37 See, notably, Articles R1125-1, R1125-3, R1125-8, R1125-10, and R1125-11 of the French Public Health Code.
38 Article L1243-1 of the French Public Health Code.
39 Article R1243-1 of the French Public Health Code, and Decree 2008-968 (Décret n° 2008-968 du 16 septembre 2008 (autorisations d'activités et de produits)).
40 Amendment of regulation 15 of the Clinical Trials Regulations, 3(c)(4B) of the Medicines for Human Use (Clinical Trials) and Blood Safety and Quality (Amendment) Regulations 2008 S.I. 2008/941.
41 Department of Health, GTAC, Agreed Decision tree, at: <http://www.dh.gov.uk/prod_consum_dh/groups/dh_digitalassets/@dh/@en/documents/digitalasset/dh_087984.pdf>.
42 Article L531-3 of the French Environment Code.

stratification of risks for clinical trials of such health products, associated with institutional flexibility. In both cases, it appears therefore that there is significant scope for discretionary decision making, under conditions of uncertainty, on a case-by-case basis by constituted groups of experts, in spite of, or at least alongside, the complex decision trees and specific national-level regulatory institutions that have been developed.

It is interesting to note that in both France and the United Kingdom, there is a specific committee in charge of the assessment of risks for particular gene therapy. The remit regarding the range of types of product of the HCB in France is wider than that of the GTAC in the United Kingdom, showing that there can be flexibility built in to the scope of apparently specifically-designed regulatory bodies, partly for the pragmatic reason of the limited available expertise and its organization, and pointing to the unstable and open-ended nature of the regulated objects.

### 3. The EU regulation for marketing authorization and follow-up of regenerative products as advanced therapy medicinal products

Commercial establishments brought under regulation of EU competition laws are brought into an EU 'harmonized' regime. On 13 November 2007, the EU adopted a *lex specialis*[43] addressing what were then legally termed advanced therapies, including TEPs, within a claimed single coherent framework in order to bridge the pre-existing regulatory gap for TEPs: Regulation (EC) No. 1394/2007 ('the Regulation on ATMP').[44] This includes a provision for a central and unique marketing authorization at the European Medicines Agency (EMA) level where a new Committee for Advanced Therapy (CAT) has been created,[45] meaning that once authorized, products may be made available throughout the EU member states without recourse to separate national marketing authorizations.

Since the adoption of the Regulation on ATMP, four types of biological medicinal products based on genes, cells, and tissues are regulated at EU level: gene therapy medicinal products (GTMP), somatic cell therapy medicinal products (CTMP), tissue-engineered products (TEPs), and combined ATMP[46] which associates a medical device with an advanced therapy. The EMA can provide an informal scientific recommendation on the

---

43 The Latin '*lex specialis*' notion comes from the legal maxim '*lex specialis derogat legi generali*'. A '*lex specialis*' is a 'law' which governs a specific subject matter. The legal maxim means that a law governing a specific subject matter overrides a law that only governs general matters. For our subject, Regulation (EC) No. 1394/2007 overrides the general EU pharmaceutical legislation.
44 OJ L324/121, 10.12.2007.
45 The Marketing Authorization is granted by the European Commission after consulting the new CAT and the Committee for Medicinal Products for Human Use within the EMA.
46 ATMP Regulation, op. cit., n. 9, Article 2.1(d).

classification of products.[47] The Regulation applies to products which correspond to the EU legal definitions and 'which are intended to be placed on the market in Member States and are *either prepared industrially or manufactured by a method involving an industrial process*'[48] (our emphasis). Unsurprisingly perhaps, there are difficulties in distinguishing which products are covered and which are not. The European Commission has tried to clarify the 'industrial process':

> This should cover, inter alia: Any 'mass production' of advanced therapy products for allogeneic use (batch production, 'off the shelf' products etc.); any advanced therapy product for autologous use (i.e. using cells/tissues from a single patient and re-implanting after manipulation into same patient) which, although being patient-specific by definition, is manufactured in accordance with a standardised and industrial process.[49]

This classificatory distinction is crucial to defining the status and responsibilities of producers of regenerative products, whether in hospitals or in the commercial sector, and was the subject of major debate amongst interested stakeholders in the negotiation of the Regulation.[50] The Regulation also strengthens 'post-authorization' requirements[51] and provides a reinforced traceability.[52]

The ATMP Regulation is directly enforceable in British and French laws, which have been modified to comply with EU law through the Medicines for Human Use (Advanced Therapy Medicinal Products and Miscellaneous Amendments) Regulations 2010 for the United Kingdom, and a new law adapting French law to the EU law adopted on 22 March 2011[53] in which Article 8 is devoted to ATMP for France.

Having described and discussed the regulation of human tissues and cells as medical materials, clinical trial aspects, and products as advanced therapies, we turn, secondly, to examine the question of body elements and products which may remain not covered by these EU-level laws. As noted in our introduction, this highlights further the question of under what circumstances is a product to be regarded as prepared by an 'industrial process,' which has direct consequences for the regimes that will apply and the commercial and safety interests of producers and consumers – companies, hospitals, physicians, and patients.

---

47 id., Article 17.
48 id., Preamble (6).
49 Commission staff working document.
50 See A. Faulkner, *Medical Technology into Healthcare and Society* (2009) ch. 8.
51 ATMP Regulation, op. cit., n. 9, Article 14.
52 id., Article 15.
53 Law 2011-302 of 22 March 2011, *J. Officiel De La République Française* (23 mars 2011) texte n° 6.

# REGENERATIVE TECHNOLOGIES NOT REGULATED BY EU LAW AS ADVANCED THERAPY MEDICINAL PRODUCTS: THE INFLUENCE OF EU LAW IN FRANCE AND THE UNITED KINGDOM

## 1. Tissues and cells for graft or transplant

In France and the United Kingdom, tissues and cells used for graft or transplant are not classified as medicinal products. They are regulated in order to comply with the EU Directives on tissues and cells outlined above.

In France, the regulation of human tissues and cells for therapeutic purposes is the same as for 'preparations' of human cell therapy. Producers must obtain an authorization from AFSSAPS, after assessment of their processes for preparation and preservation, as well as their therapeutic indications, and after advice and consent from the Biomedecine Agency.[54] Various specific decrees have been drawn up such as Decree no. 2008-968[55] which makes a distinction between institution authorization for preparation, preservation, distribution and transfer of tissues, their by-products, cells and preparations derived from cell therapy and authorization of the *processes* for such materials.[56]

In the United Kingdom, on the other hand, an establishment which wants to store human tissues and cells for human application needs a licence. Regarding the activities of procurement, testing of donor samples, processing, import and export, and distribution of tissues and cells for human application, either a licence or a third-party agreement with a licensed establishment is required. Such a licence is currently provided by the HTA in accordance with the Human Tissue (Quality and Safety for Human Application) Regulations 2007 implementing the EU Directives on tissues and cells.

## 2. The 'hospital exemption'

As noted above, the ATMP Regulation provides a 'hospital exemption'; ATMPs which are:

> prepared on a non-routine basis according to specific quality standards, and used within the same Member State in a hospital under the exclusive responsibility of a medical practitioner, in order to comply with an individual medical prescription for a custom-made product for an individual patient[57]

are not covered. Thus, manufacture under the hospital exemption must be authorized by member states. In French law, ATMPs under the hospital

---

54 Article L1243-5 of the French Public Health Code.
55 Decree 2008-968, op. cit., n. 39.
56 Article R1243-1, op. cit., n. 39, and id.
57 ATMP Regulation, op. cit., n. 9, Preamble (6) and Article 28(2).

exemption were regulated as 'preparations'. However, the new law adopted to modify French law to comply with the ATMP Regulation provides that only establishments or entities authorized by AFSSAPS, following an opinion given by the Biomedecine Agency, and pharmaceutical establishments, can prepare, preserve, distribute, and transfer ATMPs under the hospital exemption, and in accordance with good practices to be defined by the French agencies.[58] A specific Decree will be adopted to define the categories of establishments and modalities which can be authorized. In the United Kingdom, the MHRA has provided guidance on what constitutes non-routine preparation of a product.[59] Two main questions are asked: first, is it the same product, repeatedly under consideration? Second, what are the scale and frequency of the preparation of the product? The MHRA has also developed guidance on the United Kingdom's arrangements under the hospital exemption scheme.[60] This sets up specific standards, including on good manufacturing practice and quality, pharmaco-vigilance, traceability, sanctions and penalties, and requirements outside the Regulation such as labelling, package leaflet requirements, and advertising. Such guidance does not provide new legislative requirements under the hospital exemption, stipulating that NHS trusts may deal with clinical ethical issues and the GTAC could provide ethical advice.

Again, in both France and the United Kingdom, we can see the way in which the attempt to regulate therapies not intended for commercial marketing grapples with the 'boundary work' of adapting existing regimes to new laws, and the potential for uncertainty in applying classifications that make commercially consequential sectoral (hospital, company) distinctions between the actors in the regenerative therapy field.

3. *Others categories in national laws*

Separate from grafting and transplant and hospital exemption provisions, France and the United Kingdom have a range of further provisions designed to cover particular types of product or methods of preparation or production. Space prohibits a full discussion of these, but we note the following key points. First, in French law, gene therapy and xenogeneic cell and tissue engineering preparations are considered as medicinal products although they are not manufactured at an industrial scale. Establishments and processes have to be authorized by AFSSAPS. Preparations from human tissue engineering and human cell therapy fall under the regulations governing human cells and tissues rather than medicinal products, a fine distinction reflecting the particular history and basis of French law. Second, in British law, two other

---

58 See, notably, articles L4211-9, L5124-1, L5121-5 of the French Public Health Code.
59 MHRA, Annex B, Guidance on 'non-routine'.
60 MHRA, ATMPs Guidance.

regimes can be applied to medicinal products based on genes, cells or tissues, the pre-existing United Kingdom 'Specials' exemption, or the Medicines Act exemptions. These exemptions are complex and are under development at the time of writing; at first sight, they appear difficult to distinguish from the ATMP hospital exemption. The 'specials' scheme provides that to 'fulfil special needs' the provisions of the medicines Directive need not be met to respond to 'a bona fide unsolicited order ... for use by an individual patient'. For such products only a manufacturer's licence is required. Further, although in principle a product licence is required to procure, sell, supply or export a medicinal product, along with a manufacturer's licence,[61] exemptions are provided for doctors and pharmacists by the Medicines Act 1968. Such exemptions would apply to ATMPs as to other medicinal products.

Thus it appears that the regulatory work of providing for exemptions for the production and use of regenerative therapy products in the EU regime is being addressed strongly but differently in the United Kingdom and France as they wrestle with existing regimes. The different ethically-motivated emphasis between the two states reappears here, with France again relying more on a standpoint on non-commodification (shown by the refusal to qualify human cell or tissues based preparations as 'products') compared to the United Kingdom's greater concern with product licensing issues.

## DISCUSSION

Processes of classification of medical products can be seen as part of society's regulatory ordering of innovating technological sectors or zones.[62] They demonstrate change of constitutional significance in the relations between living entities, technologies, and states.[63] Society's regulatory classifications of technology, as in other arenas, have important consequences for how risks and benefits are perceived,[64] constructed, and managed; what regimes of evidence are brought into play; what private or public resources are deployed; and what characteristics shape publics' approval or concerns about the uses of the technology. The structuring work of classification is particularly striking in the socio-medical and industrial worlds of medicine and healthcare.[65]

---

61 Section 7(2) of the Medicines Act 1968.
62 A. Faulkner, 'Regulatory policy as innovation: constructing rules of engagement of a technological zone for tissue engineering in the European Union' (2009) 38 *Research Policy* 637–46.
63 S. Jasanoff, *Designs on Nature: Science and Democracy in Europe and the United States* (2005).
64 For more detail, see M.L. Flear, 'The EU's biopolitical governance of advanced therapy medicinal products' (2009) 16 *Maastricht J. of European and Comparative Law* 113–37.
65 G.C. Bowker and S.L. Star, *Sorting Things Out: Classification and its Consequences* (2000).

Classifying and formalizing practices shape new kinds of attachments that enrol persons in the performance of public practices such as medicine.[66] One of the main avowed aims of the law building that we have described here is to achieve 'legal clarity' for human tissues and cells for therapeutic use and for regenerative technologies, and there is evidence of 'commensuration' processes[67] in the regulatory work of aligning different types of medical product with each other in the legislation discussed. Yet, in contrast, the high degree of complexity in distinctions between different materials, production processes, and institutional participants of regenerative medicine remains evident from our descriptions of the key legal provisions in the EU and two member states. The hybrid, transgressive characteristics of many investigational therapies and potential products is associated with a plethora of new legal regulations and exemptions from them. Classifications can usefully be seen as maps to the world of medical technologies. Indeed, in the United Kingdom at least, the production of 'roadmaps' by regulatory authorities is a distinctive feature of their communication with the world of regenerative medicine producers and researchers. Classifications emerge closely associated with the development of experimental scientific and clinical apparatuses as well as the production of scientific facts and institutional organization. One metaphor for this is of elements being 'mangled' together, adapted to the extreme diversity of biomedical contingencies.[68]

How do we explain the detailed progress of classification and sub-classification of different materials and production and preparation processes constructing the 'material world' of regenerative medicine? Is it merely a question of the law progressively being evolved in attempts to respond to unstable, complex biologies and technologies, as we asked in our introduction? There are several answers to this question. First, from a legal point of view, given the peculiar federal nature of the EU polity, a main motive of classification comes from the principle of sharing competence between the EU and its member states. The EU can only regulate where it has competency, in this case, in cross-border movement of products (considered as goods in EU law and therefore submitted to the principle of free movement of goods) and where action at the EU level has added value in accordance with the subsidiarity principle applying in the field of public health. Member states may legislate individually on matters of national concern, such as, in the case considered here, products derived from human embryonic stem cells. This 'top-down' argument should be supplemented by a 'bottom-up' approach, that is, local forms of regulatory adjustments involving stakeholders at a national level. From a sociological point of view, one can

---

66 N. Marrres, 'The Issues Deserve More Credit: Pragmatist contributions to the study of public involvement in controversy' (2007) 37 *Social Studies of Sci.* 759–80.
67 W.N. Espeland and M.L. Stevens 'Commensuration as a social process' (1998) 24 *Ann. Rev. of Sociology* 313–43.
68 A. Pickering, *The Mangle of Practice: Time, Agency and Science* (1995).

suggest that the 'boundary work'[69] of national implementation of EU laws amounts to a form of partial 'nationalization' of biomedical therapy and biomedical 'industry' – a phenomenon that has been called 'techno-nationalism'.[70] This may be so, but from a legal perspective, the preferred interpretation would be that such boundary work is a national retention of competency by member states, which can of course result in varying priority being given to biomedicine. This boundary work of national implementation of EU law can also be seen as a bottom-up process that materializes in specific local spaces, which can be conceptualized as 'technological zones',[71] characterized by the development of common standards and assessments of objects and practices, neither clearly bounded nor necessarily corresponding to the borders of nation-states.

Secondly, the agreed high level of risks for these biological products gives rise to a high degree of stringency which requires regulatory attention to procedures and products at a very high level of technical detailing. However, the 'stringency' that is being constructed here is ambiguous. On the one hand, we see a high degree of elaboration of EU legal frameworks and national provisions, providing for exceptions and adapting existing laws and principles. But on the other hand, it is notable that the ATMP product law in particular is open-ended on many matters concerning the technical assessment of yet-to-emerge products, justified on the grounds that these cannot be foreseen. Thus, issues of, for example, clinical data requirements for ATMPs are left outside society's (parliamentary) review process for 'comitology', that is, to the technical committees of the European Medicines Agency and its consultative processes.[72]

Thirdly, and particularly for France as this discussion shows, the human materials character of these products highlights a distinctive classificatory organizing principle which has complex ramifications and is based on the strong attachment of French bioethics to respect for the autonomy and integrity of the human body and its elements, resting on upholding the ethical distinction between 'the thing and the person'. This derives from deep cultural and bioethical traditions, as well as being influenced by the well-known blood contamination scandals which reached the highest echelons of French government. Respect for the human body is expressed differently in the United Kingdom's implementation of the recent EU laws,

---

69 T.F. Gieryn, 'Boundary-work and the demarcation of science from non-science: strains and interests in professional ideologies of scientists' (1983) 48 *Am. Sociological Rev.* 781–95.
70 D. Edgerton, 'The Contradictions of Techno-Nationalism and Techno-Globalism: A Historical Perspective' (2007) 1 *New Global Studies*, at <http://www.bepress.com/ngs/vol1/iss1/art1>.
71 A. Barry, 'Technological Zones' (2006) 9 *European J. of Social Theory* 239–53; Faulkner, op. cit., n. 62.
72 A. Mahalatchimy et al., 'The European Medicines Agency: a public health European Agency?' (2012) *J. of Medicine and Law* (forthcoming).

notably in the detailed exposition of the principle of informed consent in the Human Tissue Act, whose ramifications are materialized more in social procedures than in substantive person/property boundaries. Thus the two systems are markedly different in approach, though it remains arguable as to whether one is more or less 'stringent' compared to the other.

The legal developments described here position the EU and individual member states in the global world of regenerative medicine in which stakeholders compete and collaborate.[73] In the still-evolving legal framework, we can see a mixture of measures designed to give commercial producers improved access to the European marketplace while protecting individualized biotherapies created in hospitals, and constructing stringent safety and risk management regulations for both. Having noted differences in the principles shaping the regulatory approaches of France and the United Kingdom, it is of interest to ask if there are differences between the two countries in the *implications* of the complex regulatory frameworks for biotherapy producers and users or patients in the respective healthcare systems. On the basis of this discussion, the implications appear similar in France and the United Kingdom. Indeed, the regulations across the differentiated categories of therapy are, in regulatory terms, deemed overall to enhance competitiveness for companies and foster innovation, at the same time providing, whether at EU or national level, very stringent requirements regarding the high level of safety risk. For users and patients, there are similar implications in the EU, and in France and the United Kingdom individually, for patients to access safe medicinal products.[74] Although clear differences might emerge in the means and manner of reimbursement of products, that topic is beyond the scope of the present article.

What can we conclude about comparing the configuration of the different sets of regulatory *institutions* in each society, and how they 'match', or not, their corresponding, mobile regulatory objects and processes? These issues have been discussed in relation to other biomedical technologies.[75] As noted, in France there is one primary national regulating agency for such products (either medical materials or health products including medicinal products) used for therapy, whereas in the United Kingdom three main national agencies are relevant. It could be argued that the French set-up matches the structure of the combined tissue and cells and advanced therapy framework more closely than the United Kingdom's, because the Regulation does not distinguish between different types of starting material, and this is also the case with the French national agency, although it does contain different

---

73 S. Saurugger, *Théories et concepts de l'intégration européenne* (2009).
74 A. Mahalatchimy, 'Access to advanced therapy medicinal products in the European Union: where do we stand?' (2011) 18 *European J. of Health Law* 305–17.
75 N. Brown et al., 'Regulating Hybrids: "Making a Mess" and "Cleaning Up" in Tissue Engineering and Transpecies Transplantation' (2006) 4 *Social Theory & Health* 1–24.

pathways and product categories. Conversely, the differentiated agencies in the United Kingdom could also be seen as corresponding better to the two sets of complementary EU regulations (for tissues and cells, and for medicinal products) compared to the French agency. At the beginning of this article, we posed the question of whether regulatory institutional structures might 'naturally' reflect unstable biological and technical categorizations, or whether institutional structures were relatively 'free-floating'. Our analysis here, showing the wide national variation in structures implementing the same EU-level laws designed and adapted to cover the same field, supports the latter interpretation: regulatory institutional structures here have a high degree of flexibility in relation to the 'underlying' processes of material innovation, though they strain against existing, inherited legislations. Indeed, it would be more accurate to understand this not as 'implementation' but rather as a process of adaptation of EU laws, in which influence flows from EU to national jurisdictions and vice versa, and, in sociological terms, one of co-production between regulatory institutions and regulated 'objects'. 'Local' expectations, fears, and promises surrounding regenerative medicine's objects undoubtedly shape the institutional regulation of material innovation, possibly to a greater degree than the scientific data that accompany those objects.[76] Governance activity has a constructive and shaping action as well as standardizing some material practices. The dynamics of the emerging medical world thus perform a complex web of codified legal narratives.[77]

This review necessarily leaves some loose ends. Although a legal framework has been created, some scientific and technological innovations escape its categories. At the same time, some aspects of the regulation at EU level are deliberately left loose, open to new technical developments. The two countries discussed here attempt to reconcile new rules with pre-existing, institutionalized pharmaceutical and human tissue related principles – the 'inherited regulatory environment'.[78] Perhaps most notable in terms of the materialization of law, in conclusion, is that alongside the elaborate construction of a new regulatory framework at EU level and its adaptation in national regimes, we also see that each new regenerative medicine product will have to receive a unique scientific expert assessment before being made available as a product in the healthcare system. It may be that the operation of discretion by committees of selected specialists considering innovations

---

76 N. Brown, 'Shifting Tenses – From Regimes of Truth to Regimes of Hope?' (2007) 13 *Configurations* 331–55.
77 A. Faulkner, *How Law Makes Technoscience: The Shaping of Expectations, Actors and Accountabilities in Regenerative Medicine in Europe.* (2010) CSSP Electronic Working Paper No.1, at <http://www.jnu.ac.in/Academics/Schools/SchoolOfSocialSciences/CSSP/CSSP-EWPS-1.pdf>.
78 Compare E. Stokes, 'Nanotechnology and the Products of Inherited Regulation' in this volume, pp. 93–112.

on a case-by-case basis will be the over-riding feature for deciding the regulatory route that new products and their developers will have to take. Experts' legitimacy here takes the form of institutional legitimacy via their formal affiliation to the European Medicines Agency as a mandated European Agency, and the technical legitimacy afforded by specific recognized skills. Thus, we conclude by pointing to the fact that, in spite of the regulatory reach of a complex and highly detailed new legal framework, a tailored case-by-case approach to deploying technical expertise nevertheless is required in regulating the fast-moving, innovative arena of regenerative medicine.

# Bodies of Science and Law: Forensic DNA Profiling, Biological Bodies, and Biopower

VICTOR TOOM\*

*How is jurisdiction transferred from an individual's biological body to agents of power such as the police, public prosecutors, and the judiciary, and what happens to these biological bodies when transformed from private into public objects? These questions are examined by analysing bodies situated at the intersection of science and law. More specifically, the transformation of 'private bodies' into 'public bodies' is analysed by going into the details of forensic DNA profiling in the Dutch jurisdiction. It will be argued that various 'forensic genetic practices' enact different 'forensic genetic bodies'. These enacted forensic genetic bodies are connected with various infringements of civil rights, which become articulated in exploring these forensic genetic bodies' 'normative registers'.*

## INTRODUCTION

In recent years, sociological studies of forensic science have gained momentum. In particular, forensic DNA profiling has received a considerable degree of scrutiny from scholars drawing upon approaches from science and technology studies.[1] These studies have addressed various issues including

---

\* *Northumbria University Centre for Forensic Science (NUCFS), Northumberland Building, Newcastle upon Tyne NE1 8ST, England*
victor.toom@northumbria.ac.uk

Alex Faulkner and Christopher Lawless gave me the opportunity to contribute to the volume, reviewed and commented on earlier drafts, and corrected the English. Valuable comments were also provided by an anonymous reviewer of the *Journal of Law and Society*, Bart van der Sloot, and Katharina Paul. I want to thank them all for their support and for stimulating me to proceed.

1 J. Aronson, *Genetic Witness: Science, Law, and Controversy in the Making of DNA Profiling* (2007); S. Jasanoff, 'The eye of everyman: witnessing DNA in the Simpson trial' (1998) 28 *Social Studies of Science* 713; M. Lynch, 'The discursive production of uncertainty: the OJ Simpson "Dream Team" and the sociology of knowledge machine' (1998) 28 *Social Studies of Science* 829–68; M. Lynch, S. Cole, R.

the most plausible interpretation of DNA evidence, questions regarding the most credible expert witness, and how rules of evidence contribute to the scope and content of DNA evidence. In addition to this research, recent studies on forensic DNA profiling have focused on the application of forensic DNA profiling in criminal investigations and the (inter)national governance of DNA databases.[2] Comparatively little attention has been devoted to the relation between forensic DNA profiling, biological bodies, and bodily samples.[3] It is this nexus which is addressed in the present contribution. More specifically, it will analyse how biological bodies and bodily samples are fitted into forensic DNA practices situated at the intersection of (forensic) science and (criminal) law. The article analyses forensic DNA profiling in the Netherlands, and how, as a result of the continually advancing applications of DNA profiling, biological bodies and bodily samples become ever more important markers for the pursuit of judicial truth, administering justice, and crime control.

I elucidate the scientific and legal mechanisms which have evolved for transferring jurisdiction over an individual's biological body to agents of judicial power, including police, public prosecutor, and the judiciary. I identify these by first analysing an example of a medical case and secondly, by introducing forensic DNA profiling by way of a simplified example. I then articulate the identified mechanism by briefly examining the work of Michel Foucault and Giorgio Agamben on biopower. In the next section, I analyse Dutch forensic genetic practices, and how biological bodies and bodily samples are fitted into those practices, by tracing how forensic DNA technologies and laws to regulate them have co-evolved over the past twenty years. I demonstrate that different forensic genetic practices emerged – one directed at individuals, another at populations – and that these different practices come with different normative issues. In the concluding section, I summarize the findings and draw overarching lessons.

## IDENTIFYING THE MECHANISM

Most people would intuitively accept the proposition that 'I own my body'. Such a statement underscores the individual mastery over a particular body and delineates the private from the public. Yet, if you are under suspicion or

---

McNally, and K. Jordan, *Truth Machine. The Contentious History of DNA Fingerprinting* (2008).

2 R. Hindmarsh and B. Prainsack (eds.), *Genetic Suspects. Global Governance of Forensic DNA Profiling and Databasing* (2010); B. Prainsack and V. Toom, 'The Prüm regime: Situated dis/empowerment in transnational DNA profile exchange' (2010) 50 *Brit. J. of Criminology* 1117–35; R. Williams and P. Johnson, *Genetic Policing. The Use of DNA in Criminal Investigations* (2008).

3 An exception is: C. Kruse, 'Forensic evidence: materializing bodies, materializing crimes' (2010) 17 *European J. of Women's Studies* 363–77.

convicted for a crime, or diagnosed with a specific disease or mental condition, the statement may no longer apply; you can be arrested, searched, taken into custody, hospitalized or forced to take medications. The body that once was yours becomes an object of incrimination, incarceration, care or treatment. When jurisdiction over a body is transferred from an individual to agents of power like police and medics, those bodies transform from 'private bodies' into 'public bodies'.

An analysis regarding the transformation of private into public bodies is provided by American social geographer David Delaney.[4] He describes a case where an inmate is diagnosed with a mental disorder. The patient/ prisoner receives pharmaceutical treatment, yet withdraws consent for treatment after he experiences deleterious effects caused by the medication. Bypassing his will, the state legally continues to administer the drugs to the inmate and hence he becomes the 'unwilling recipient of a sort of "synthetic sanity".'[5] In other words, his body is put under restraint, 'his skin and muscles are penetrated by the state apparatus of the syringe, his circulatory and nervous systems are colonized by the authorities aided by pharmaceutical corporations.'[6] In this act of putting his body under restraint – that is, when a private body is transformed into a public body – several normative and legal issues arise: individual consent and autonomy are bypassed and his right to an inviolable body is breached. His body is not 'his' anymore. Or, as Delaney puts it, the inmate's body becomes 'a material slab, a zone between his self and the outer institutional environment.'[7]

Before any individual can be made a ward of court, evidence that he is not able to take care of himself should be provided. In the example provided by Delaney, it can be assumed that the inmate was tested both mentally and physically (using the *Diagnostic and Statistical Manual of Mental Disorders*, DSM), observed and interviewed, and that all collected data was evaluated. These results are compared with other scientific data, like fMRI scans or neurochemically tested samples. Individuals can only be made wards of court if the results point to the same mental condition. Hence, scientific evidence warrants the legal decision of transferring jurisdiction over a body from the individual to an authority – such body *becomes* constituted as a *body of science and law*.

The criminal justice system is a domain where private bodies are rendered into public bodies routinely. Several mechanisms are in place to achieve this, the most straightforward being the arrest or imprisonment of individuals. Such public bodies can be regarded as 'bodies of law' as legal mechanisms are in place to warrant arrests, and so on. Bodies of science and law have

---

4 D. Delaney, 'Making nature/marking humans: law as a site of (cultural) production' (2001) 91 *Annals of the Association of Am. Geographers* 487–503.
5 id., p. 499.
6 id.
7 id.

been around since as long ago as the nineteenth century, when scientific methods, like dactyloscopy (fingerprinting) and anthropometry (biometrics) were already being applied to make representations of individuals and their bodies.[8]

Since the introduction of forensic DNA profiling in criminal justice systems, bodies of science and law have changed considerably, not least because the latter technology analyses bodily samples. The example below will introduce forensic DNA profiling and its intimate relationship with bodies and bodily samples:

> A break-in was reported by a witness who saw a man smashing in a window. After the police had arrived, a crime scene examiner collected a piece of glass and secured a bloodstain probably originating with the burglar. The blood trace was submitted to a forensic laboratory where a DNA profile was obtained and subsequently uploaded to the national DNA database. In the meantime, the police arrested an individual on suspicion of having committed the burglary. He was asked to provide a biological reference sample, which he refused to do. He was then physically forced by two police officers, who opened his mouth to take a buccal swab. The sample was processed at the forensic laboratory and it matched the DNA crime scene sample. Such a match is usually expressed as a statistical number (the random match probability), stating that the chance that someone in the population at large would have the same DNA profile as less than one in a billion.[9]

Agents of power like the police, public prosecutor, and judge are, since the introduction of forensic DNA profiling in criminal justice systems, advancing further into personal spheres, thereby rendering the personal into public objects.[10] The radical shift that has occurred is that crime investigation and criminal litigation have become intimately connected with body samples and the production of genetic knowledge about those samples, and hence its originators. It is in this capacity that the mechanism that enables forensic DNA profiling resembles key aspects of Michel Foucault's analysis of 'biopower'. First, genetic and subsequent digital representations of bodies are produced, enabling the comparability of bodily samples originating from a subject and bodily traces collected at crime scenes. Such representations contain *knowledge* of the originator's body that goes beyond 'the science of its functioning'.[11] Second, authorities involved in criminal investigation, who have gained procedural powers to collect biological samples from subjects, process the samples into DNA profiles, and store samples and

---

8 S.A. Cole, *Suspect Identities. A History of Fingerprinting and Criminal Identification* (2001).
9 The example is derived from V. Toom, *Dragers van Waarheid. Twintig Jaar Forensisch DNA-onderzoek in Nederland* ('Carriers of Truth. Twenty Years of Forensic DNA Profiling in the Netherlands') (2011).
10 Alcohol and drug tests are also dependent on the use of body samples. Yet, police in the Netherlands have never had legal powers physically to compel individuals to supply body samples.
11 M. Foucault, *Discipline and Punish. The Birth of the Prison* (1977) 26.

profiles in biobanks and databases for many years. So, the authorities have gained *mastery* over those bodies and body parts for a specific amount of time which 'is more than the ability to conquer them'.[12] Mastery over bodies and the ability to produce knowledge about them are central to Foucault's analyses regarding the workings of power and its relationships with bodies.[13]

Yet, to appreciate what goes on in forensic DNA profiling, and to be able to analyse the normative consequences of the routine criminal justice mechanism that renders private bodies into public bodies, it is essential to enquire briefly into political philosophy and legal theory. Everybody in a democratic state of law is entitled to civil rights. Freedom of speech, religion, association, and opinion are some examples. Such rights, especially in continental liberal democracies, are laid down in a system of rights usually referred to as the Constitution. Rights of the 'self' are, in the context of forensic DNA profiling, regarded as personal lives and the integrity of individual bodies. In order to protect these realms against the power of the state and its institutions, like the police, the Office of Public Prosecution, and the judiciary, individuals, their bodies and personal lives are assigned civil rights.

More specifically, turning to the Netherlands as a case study, personal lives and individual bodies are protected by articles 10 and 11 of the Dutch Constitution, which states that everyone shall 'have the right to respect for his privacy' and 'have the right to inviolability of his person'.[14] These rights are not absolute rights as they can be curtailed according to 'restrictions laid down by or pursuant to an Act of Parliament'.[15] In other words, personal lives and individual bodies can be violated by authorities involved in criminal investigation when conditions described in the Code of Criminal Procedures are met. With regard to Dutch DNA profiling, the Code of Criminal Procedures delineates when privacy and the right to an inviolable body do not apply to an individual, and typically include suspects and convicted offenders (see further below). The bodies and personal lives of these categories of individuals are excluded from Articles 10 and 11 of the Dutch Constitution. Their bodies and bodily samples are, echoing Giorgio Agamben's influential work on contemporary mechanisms of biopower, legally in a 'state of exception'.[16]

---

12 id.
13 id.
14 Ministry of the Interior and Kingdom Relations, *The Constitution of the Kingdom of the Netherlands* (2002). Both Articles are in concordance with Article 8 of the European Convention on Human Rights and Article 12 of the Universal Declaration on Human Rights. In Anglo-American jurisdictions, violations of the body are sometimes defined as an infringement of 'spatial' privacy, whereas personal information is usually called 'informational' privacy: see G. Laurie, *Genetic Privacy. A Challenge to Medico-Legal Norms* (2002).
15 Ministry of the Interior and Kingdom Relations, id., p. 6.
16 G. Agamben, *Homo Sacer: Sovereign Power and Bare Life* (1998).

This section has identified the mechanisms of how agents of power progress into personal spheres when forensic DNA profiling is applied. The mechanism works on three different levels. First, authorities involved in criminal investigation gain mastery over individual bodies. Second, this mastery can only be gained when individuals and their bodies are stripped of entitlements to civil rights. Third, only when these conditions are met can forensic genetic knowledge about those bodies be produced. This mechanism renders private bodies into public bodies, constituted both in science and in law. To delineate these produced public bodies from other conceivable public bodies (like the one described by Delaney and other biometric bodies), I will call these bodies 'forensic genetic bodies'. Having established this trend, we will next analyse details of Dutch forensic genetic practices by tracing the co-evolution of forensic DNA technologies and laws to regulate them.[17] The aim of the analysis presented in the next section is to establish empirically how forensic genetic bodies have been conscripted into the criminal justice system, and what the normative consequences of this are.

## FORENSIC GENETIC PRACTICES IN THE NETHERLANDS

This section traces developments in the Dutch jurisdiction regarding forensic DNA profiling and criminal law to regulate it. It will be argued that two related yet different 'forensic genetic practices' have emerged: one directed at individuals, the other at populations. My analysis does not start with a fixed idea of what forensic DNA profiling *is*, but what it *becomes* when it is used in forensic genetic practice – when it is being practiced in a 'situated event'.[18] In addition to analysing these practices, the aim of this section is to establish empirically how private bodies are transformed into forensic genetic bodies and to articulate the variety of normative (or political) concerns involved. These different sets of normative concerns I shall refer to as 'normative registers'. This reflects my aim of cataloguing the normative content of each forensic genetic practice and how bodies and bodily samples are incorporated into that practice. I use the work of Dutch philosopher Annemarie Mol as a guide to practices, bodies, and normative registers. She proposes using the concept of 'enactment' for the analysis of the effects of interacting objects, logics, and practices.[19] Or, as she states, enactment suggests 'that activities take place – but leaves the actors vague. It also

---

17 This process has been termed 'biolegality', see M. Lynch and R. McNally, 'Forensic DNA Databases: The Co-Production of Law and Surveillance Technologies' in *Handbook of Genetics and Society: Mapping the New Genomic Era*, eds. P. Atkinson, P. Glasner, and M. Lock (2009) 283–301.
18 A. Mol, 'Actor network theory: sensitive terms and enduring tensions' (2011) 50 *Kölner Zeitschrift für Soziologie und Sozialpsychologie* 253–69.
19 A. Mol, *The Body Multiple. Ontology in Medical Practice* (2002).

suggests that in the act, and only then and there, something *is* – being enacted.'[20] In addition, enactment neatly resonates with the legal and political system that enacts laws. Science and law are the most prominent actors in my analysis here. This, however, does not mean that forensic genetic practices develop or become enacted outside of the influence of people, 'the social' or institutions. Below, I first analyse forensic genetic practices aimed at individuals, how they enact 'known bodies', and how these known bodies have become associated with a specific normative register. Attention will then be devoted to new forensic genetic technologies aimed at determining personal external visible characteristics from 'unknown' human originators. I argue that this practice is dependent on another enacted forensic genetic body and is intimately connected with the organization of DNA dragnets, and demonstrate how bodies are fitted into criminal investigation in novel ways. As such, it generates another normative register which is characterized by its clustering capacities.

1. *Individualizing forensic genetic practices*

Like many other 'Western' jurisdictions, the Dutch judiciary accepted DNA evidence in the late 1980s.[21] The first DNA case in the Netherlands regarded a suspect who was convicted of rape by a trial judge yet proved to be innocent on appeal after volunteering a DNA sample. It was, however, legally impossible to use force against individuals to obtain reference samples as article 11 of the aforementioned Dutch Constitution applied to blood 'under the skin' and saliva 'in the mouth'. Many Members of Parliament realized that forensic DNA profiling could be a valuable addition to the arsenal of existing technologies for individualization. It was therefore relatively soon decided to make mandatory DNA profiling legally possible by redistributing jurisdiction over bodies and biological samples therein from individuals to the judiciary when specific conditions were met. But before such measures could be enacted, technical, legal, and ethical issues had to be considered.

Forensic DNA technologies as applied until the mid 1990s were infamous because of several drawbacks, including susceptibility to false interpretation and contamination, the lack of any rigorous standards for the production and interpretation of DNA profiles, high costs, and the duration of the cycle of analysis.[22] In addition to these problems, forensic DNA technologies were

---

20 id., p. 33 (emphasis in original).
21 Hindmarsh and Prainsack, op. cit., n. 2.
22 For a full discussion, see E.S. Lander, 'DNA fingerprinting on trial' (1989) 339 *Nature* 501–5; E.S. Lander, 'DNA fingerprinting: science, law, and the ultimate identifier' in *The Code of Codes. Scientific and Social Issues in the Human Genome Project*, eds. D.J. Kevles and L. Hood (1992). For STS analyses on technical, scientific, and legal issues, see Lynch et al., op. cit., n. 1; Aronson, op. cit., n. 1.

dependent on samples containing high-quantity, non-degraded DNA typically available in blood and semen, which linked forensic DNA typing to severe and violent crimes like rape and murder. Obtaining blood samples from suspects was consequently an 'obligatory point of passage' for reliable DNA profiling.[23] Yet, using force to obtain blood samples was considered to be a drastic infringement of a suspect's right to an inviolable body.

All these issues were considered when a first bill was drafted to govern forensic DNA profiling in the Netherlands, which resulted in the 1994 Forensic DNA Profiling Act.[24] This piece of legislation rendered it impossible for certain classes of suspects to appeal to their right to an inviolable body; high thresholds were implemented before body samples could be taken without consent. First, only suspects who were under suspicion of having committed a crime with a liability of eight years or more imprisonment (that is, sex crimes, homicide) could be forced to provide a DNA sample. And second, DNA profiling of a suspect's biological sample could only be ordered by a representative of the judiciary, the investigating judge. It is for these two reasons that forensic DNA profiling initially became enacted as an evidentiary practice for severe and violent crimes.

An invention awarded the Nobel Prize for Chemistry in 1993 would come to intervene in evidentiary practice for severe and violent crimes. This technique, called the polymerase chain reaction (PCR), allowed for multiplying biological samples *in vitro* and subsequently detached forensic genetic technologies from large blood and semen stains. It contributed importantly, first, to the possibility of analysing DNA traces present on mundane physical objects (cigarette butts, glasses, garments) often collected at scenes of less serious crime (for example, burglary, car theft) and, second, enabled the production of DNA profiles obtained from buccal swabs instead of blood.

Geneticists started developing new forensic genetic typing systems in the early and mid-1990s.[25] These systems combined PCR with so-called 'short tandem repeats' (STRs). STRs had several advantages compared with the forensic DNA techniques applied earlier. The first was that the production of DNA profiles became standardized and partly computerized, relegating interpretive problems regarding DNA profiles to the background. Various

---

23 For the term 'obligatory point of passage', see M. Callon, 'Some Elements of a Sociology of Translation: Domestication of the Scallops and the Fishermen of St. Brieuc Bay' in *Power, Action and Belief. A New Sociology of Knowledge?*, ed. J. Law (1986) 196–233.

24 A. M'charek, 'Silent witness, articulate collectives: DNA evidence and the inference of visible traits' (2008) 22 *Bioethics* 519–28; V. Toom, 'DNA Fingerprinting and the Right to Inviolability of the Body and Bodily Integrity in the Netherlands: Convincing Evidence and Proliferating Body Parts' (2006) 2 *Genomics, Society and Policy* 64–74.

25 R. Sparkes et al., 'The validation of a 7-locus multiplex STR test for use in forensic casework. (I) Mixtures, ageing, degradation and species studies' (1996) 109 *International J. of Legal Medicine* 186–94.

STR markers were combined in so-called multiplexes; consequently several STRs could be determined in one reaction, a second advantage, contributing greatly to the speed of forensic DNA profiling. Less interpretation combined with a faster throughput led to a third advantage: DNA profiling became cheaper. Last but not least, STRs have a numerical format and hence allowed for digitally storing DNA profiles in databases.[26]

In 2001, an amendment to lower the thresholds for mandatory DNA analysis was added to the Forensic DNA Profiling Act. An important justification for the expansion of forensic DNA profiling was the less severe violation of a body when a saliva sample is obtained (vis-à-vis a blood sample). Mandatory DNA typing became legally possible when someone was under suspicion of having committed a crime with liability of four years of imprisonment (for example, burglary, car theft). The 2001 amendment included measures to facilitate the expected increase in forensic DNA typing, two of which were important contributions to newly enacted forensic genetic practice.[27] First, in addition to the investigating judge, responsibilities for ordering mandatory DNA analysis were also given to the leader of the criminal investigation, the public prosecutor. As such, DNA profiling was not only to be considered criminal evidence, but also became a tool for criminal investigation. Secondly, competence for obtaining biological saliva samples was extended to police officers if the suspect consented in providing a sample. When force had to be applied, physicians were to take the samples. The scope and application of forensic genetic practices as the result of interactions between science and law were thus extended from severe and violent crimes to volume crimes, and from evidence in criminal proceeding to investigative leads in criminal investigation.

The 1994 Act and its 2001 amendment both focused on suspects. A new category of individuals was conscripted into forensic genetic practices when the DNA Convicted Persons Act came into force in 2005. This law applies to the category of convicted offenders (adults and juveniles) for crimes with a maximum liability of four years imprisonment or more. To prevent the Netherlands Forensic Institute becoming overloaded with biological samples to be DNA typed, it was decided to enact the law in two stages. The first came into force in 2005 and applied to persons convicted for crimes attracting a penalty of six years or more imprisonment. Since May 2010, everyone convicted of crimes specified in the 2005 law is DNA typed.

2. *Known bodies and DNA hunts*

I have stated that my interest in analysing forensic genetic practices is regarding the enactment of bodies of (forensic) science and (criminal) law,

---

26 Williams and Johnson, op. cit., n. 2.
27 V. Toom, 'Inquisitorial forensic DNA profiling in the Netherlands and the expansion of the forensic genetic body' in Hindmarsh and Prainsack, op. cit., n. 2.

the forensic genetic body. A typical forensic genetic body becomes enacted when an individual is either suspected of having committed a crime or when she or he is convicted of a crime (both punishable by four years or more imprisonment). These bodies are 'known' by authorities in two different modes. First, police and prosecuting authorities know these bodies since they have been identified. Second, genetic knowledge will be produced from these bodies, which is the second mode of knowing. It is for these two reasons that I will refer to these enacted forensic genetic bodies as 'known bodies'. This raises the question how many known bodies have been enacted in the Netherlands.

In September 2011, it was announced that more than 124,000 DNA profiles from known individuals were uploaded (from a population of approximate 17 million) onto the Dutch DNA database.[28] The amount of DNA profiles uploaded to the DNA database can be interpreted as a proxy for the number of known bodies. When a sample is obtained, the right to an inviolable body (Article 11 of the Constitution) is necessarily violated. Since genomic samples contain all the genetic 'information' of an individual, privacy (Article 10 of the Constitution) is also at stake. A majority view on the trends I am presenting here is that these measures came in after restrictions were approved by an Act of Parliament; that many legal measures are in place that order forensic DNA profiling and warrant its proper use; that 'known bodies' are most often real criminals; that criminals are less entitled to certain civic rights in open democratic societies; and that the balance between the safety of the public at large and the infringements of individual rights is maintained. Yet, what escapes the attention of many policy workers, stakeholders, and commentators is how these known bodies interfere with other legal principles, as the next contemporary example demonstrates.[29]

The DNA Convicted Criminals Act prescribes that every convicted offender, when she or he is convicted for a crime with a liability of four years of imprisonment or more, will be DNA typed. Such crimes include serious and violent crimes (sex crimes, homicide), volume crimes (burglary, car theft) and more petty crimes (shoplifting, possession of certain stolen goods, embezzlement). The bodies of offenders convicted for any of these crimes will be enacted as known bodies. There are two routes for gathering cellular material for forensic DNA analysis. The first is a jail sentence – usually for

---

28 The Dutch DNA database was established in 1997. For an actual overview of uploaded DNA profiles, see the website maintained by the Custodian of the Dutch DNA database, at <www.dnasporen.nl>.
29 By policy workers, I refer to ministers and their staff, Members of Parliament, and other stakeholders involved in the criminal justice system such as the police and the Office of Public Prosecution. Among the commentators favouring wider applications for forensic DNA typing are communitarian philosophers: for example, see A. Etzioni, 'DNA Tests and Databases in Criminal Justice: Individual Rights and the Common Good' in *DNA and the Criminal Justice System: The Technology of Justice*, ed. D. Lazer (2004) 197–224.

more serious crimes, but also categories of volume crime like burglary. When someone is incarcerated, there is ample opportunity for the authorities to obtain a reference sample. Offenders convicted of more petty crimes will often receive a suspended sentence, probation or community service. Since everyone shall 'be treated equally in equal circumstance', samples from these individuals are DNA typed too.[30] Non-incarcerated convicted offenders receive a letter from the public prosecutor stating that they are obliged to visit a 'DNA consultant' at a local police station, when a reference sample will be taken for DNA analysis. When convicted offenders do not show up, the police will 'collect' them at their residential addresses during weekends, bank holidays or soccer matches.[31] These convicted offenders are arrested and taken to the police station to obtain a reference sample. Hence, to put it provocatively, the police organize DNA hunts to gather biological samples from individuals convicted for crimes like shoplifting, being in possession of stolen goods or embezzlement.

DNA hunts on individuals convicted for petty crimes, therefore, are warranted by Dutch legislation. Yet one wonders whether such measures correspond to Article 8 of the European Convention on Human Rights. Article 8 provides mechanisms for respecting private life, but can be violated when it is regarded as proportionate in a democratic society and 'in the interests of national security, public safety or ... for the prevention of disorder or crime'.[32] But does such collection from non-incarcerated convicted offenders contribute to the national security or public safety? Will such DNA hunts prevent future crimes? In addition, it should be noted that individuals are 'collected', taken into custody, and brought to a police station in order to take a body sample for DNA analysis. Should this be regarded as a proportionate measure, given that these individuals are often convicted of petty crimes, and then stigmatized when family members and neighbours witness their 'collection' by police officers? These questions become even more urgent since more than 10 per cent of non-incarcerated individuals are under 18 years and hence are subject to the Dutch system of youth justice, yet their profiles are retained long after they legally become adults. This is 'inconsistent with the special consideration ... in the way children (especially younger children) are dealt with in the criminal justice system.'[33] Infor-

30 Article 1 of the Dutch Constitution, see Ministry of the Interior and Kingdom Relations, op. cit., n. 14.
31 See, for example, Politie Brabant Zuid-Oost, 'Politie haalt 90 veroordeelden op ten behoeven van DNA-afname' ('Police collects 90 convicted offenders for DNA research') press release, 27 May 2009; '33 personen aangehouden voor verplichte DNA-afname' ('33 persons arrested for mandatory DNA research') *Limburgse Courant*, 6 May 2010.
32 Council of Europe, Convention for the Protection of Human Rights and Fundamental Freedoms (1950) Article 8.
33 M. Levitt and F. Tomasini, 'Bar-coded children: an exploration of issues around the inclusion of children on the England and Wales National DNA database' (2006) 2 *Genomics, Society and Policy* 41–56, at 52.

mation provided by the police indicates that approximate 40 per cent of the convicted offenders do not present themselves to the DNA consultants.[34] On top of that, DNA hunts require substantial officer resources and fiscal investments in police forces. Hence, DNA hunts may, paradoxically, proceed at the expense of more traditional policing tasks, like surveillance and the prevention of criminal activities. Put differently, organizing a DNA hunt with the aim of collecting samples from non-incarcerated convicted offenders may demonstrate one way in which the police are working hard to make society safer, yet, arguably, this does not contribute to a safer society.

### 3. *Clustering forensic genetic practices*

Although DNA profiling is often considered the most important forensic breakthrough since the application of fingerprints in the late-nineteeth century,[35] individualizing DNA profiles aimed at comparing traces with reference profiles can be considered 'traditional' forensic genetic technology.[36] Recent genetic insight enables new applications for forensic DNA profiling, for instance, the examination of biological crime-scene samples to predict external visible characteristics – like geographical descent, sex, hair and eye colour, and age – of an unknown originator.[37] Such a technique was applied for the first time in the Dutch jurisdiction when 16-year-old Marianne Vaatstra was found raped and murdered in a pasture on the morning of 1 May 1999. Apart from an abundance of DNA traces, no other crime-related facts or clues were particularly useful for the criminal investigation – as a result, the case remains unsolved. In an attempt to find the murderer, genetic technologies for predicting external visible characteristics were applied, and suggested that the perpetrator originated from north-western Europe.

Although this information aided the criminal investigation, its articulation demonstrated a legal problem: the 1994 Forensic DNA Typing Act defined DNA profiling as *exclusively* aimed at comparing DNA profiles. In other words, it was legally forbidden to use genetic technologies to determine external visible characteristics, unless the law were to be amended. Consequently, in 2003, the Law on External Visible Personal Characteristics was enacted and it did regulate the determination of external visible characteristics of 'unknown' suspects by forensic genetic methods. It currently

---

34 *Limburgse Courant*, op. cit., n. 31.
35 Cole, op. cit., n. 8; S. Jasanoff, 'Foreword' in Hindmarsh and Prainsack, op. cit., n. 2.
36 B.-J. Koops and M. Schellekens, 'Forensic DNA phenotyping: regulatory issues' (2008) 9 *Colombia Science and Technology Law Rev.* 158–202.
37 M. Kayser and P.M. Schneider, 'DNA-based prediction of human externally visible characteristics in forensics: motivations, scientific challenges, and ethical considerations' (2009) 3 *Forensic Science International: Genetics* 154–61; M. Kayser and P. De Knijff, 'Improving human forensics through advances in genetics, genomics and molecular biology' (2011) 12 *Nature Rev.: Genetics* 179–92.

governs two physical traits: sex and race. The law was deliberately designed as 'window-case legislation' to enable future physical traits to be included in the law, and to enable further genetic research for these purposes (see further below). As a result, an attempt to add a third external visible characteristic (eye colour) to the law is currently pending in Parliament.

Making predictions about external visible characteristics of the unknown suspect may be useful for police investigations when the 'usual suspects' (family, friends, partners, acquaintances) are excluded as the possible perpetrator. This also occurred in the Marianne Vaatstra case. The novelty of forensic genetic technologies aimed at predicting external visible characteristics of unknown suspects is that, for policing purposes, it groups together non-suspected individuals who look similar. In other words, individuals who have similar external visible characteristics (for example, male, brown eyes, European ancestry) become targeted as 'interesting' subjects for further investigation, not so much by facts and circumstances derived from the crime, but by means of scientific methods and genetic insights – similar-looking individuals become clustered into a 'suspect population'.[38] Each non-suspected member of this scientifically produced suspect population should then be excluded as the possible perpetrator through investigation. One method aimed at the exclusion of interesting subjects is a so-called DNA mass-screening or DNA dragnet.[39] DNA dragnets have been deployed in serious criminal investigations in various jurisdictions (for example, the United States, the United Kingdom, the Netherlands, and Germany) where the police have run out of obvious suspects. Reportedly, sometimes more than a thousand individuals were requested to deliver a sample, mostly with limited or no success – the Marianne Vaatstra case is an example.[40] DNA dragnets are therefore considered to be expensive and inefficient. Yet, if more genetic knowledge regarding the physical appearance of an unknown originator should become available, and if such knowledge were more robust, then the efficacy of DNA dragnets might – in theory – increase.

Above, I described how new forensic genetic techniques aimed at inferring external visible characteristics of an unknown originator have been available and applied in criminal investigation. It was argued that this kind

---

38 S.A Cole and M. Lynch, 'The social and legal construction of suspects' (2006) 2 *Annual Rev. of Law and Social Science* 39–60.
39 id.; M'charek, op. cit., n. 24; A. M'charek, V. Toom, and B. Prainsack, 'Bracketing off population does not advance ethical reflection on EVCs: A reply to Kayser and Schneider' (2012) 6 *Forensic Science International: Genetics* e16–e17; see, also, M. Kayser and P. Schneider, 'Reply to "Bracketing off population does not advance ethical reflections on EVCs: A reply to Kayser and Schneider" by A. M'charek, V. Toom and B. Prainsack' (2012) 6 *Forensic Science International: Genetics* e18–e19.
40 See, also, S. Krimsky and T. Simoncelli, *Genetic Justice. DNA Databanks, Criminal Investigations, and Civil Liberties* (2011); H. Washington, 'Base assumptions? Racial aspects of US DNA forensics' in Hindmarsh and Prainsack, op. cit., n. 2.

of knowledge clusters similar-looking yet non-suspected individuals into suspect populations. These suspect populations can become the object of DNA dragnets. Although DNA dragnets are regarded as expensive and inefficient, it is, in principle, possible to raise effectiveness and lower fiscal burdens if predictions on external visible characteristics become more precise and robust. This brings us to the topic of genetic research with the aim of *determining* external visible characteristics. In the case considered here, I am interested in research conducted on biological samples originating in the Dutch forensic DNA database. I interpret this research as being enabled by another form of enacted forensic genetic body, elaborated below.

## 4. Sample/ID packages, scientific research, and DNA dragnets

A DNA profile represents a known body and, as such, stands proxy for an identified individual. In forensic practices, it is essential that a known body and a DNA profile remain connected. The method to achieve this connection is usually called the 'chain of custody', which is an administrative method that refers to procedures for 'collecting, transporting and handling legally significant material'.[41] In practice, the chain of custody consists, among other things, of paperwork, administration, stickers, and bar codes to guarantee that a biological sample obtained from a known body keeps on referring to the identity of the originator. Hence, a package containing a sample and information about the identity (ID) of the originator is created and may be called a 'sample/ID package'.[42] Authorities involved in crime investigation gain mastery over these sample/ID packages by way of their exemption from civil rights, which enables authorities to produce knowledge from those sample/ID packages, and hence can be considered another form of enacted forensic genetic body.

Sample/ID packages are retained in biobanks and enact the third form of forensic genetic body that I identify: the 'sample/ID bank'. This biobank represents the population of convicted offenders in the Netherlands. The Dutch DNA database is governed by the Personal Data Protection Act, and allows for scientific research to be conducted with the *information* that it governs.[43] The Personal Data Protection Act applies not only to the (digital) DNA database and the profiles it contains, but also to the retained (biological) sample/ID packages and sample/ID bank – all these objects are

---

41 Lynch et al., op. cit., n. 1, p. 114.
42 Toom, op. cit., n. 24.
43 Staatsblad van het Koninkrijk der Nederlanden, Wet van 6 juli 2000, houdende regels inzake de bescherming van persoonsgegevens (Wet bescherming persoonsgegevens) ('Law of 6 July 2000 regarding rules for protecting personal data (Personal Data Protection Act)') (2000) 302, 1–25.

legally considered to be *information*.⁴⁴ As the DNA database and its biological twin brother represent the genetic diversity of individuals convicted in the Netherlands, it provides geneticists with ample opportunity to search for, and validate, genetic markers for new external visible characteristics to be used in criminal investigation, amongst other matters.

In 2008, the Minister of Justice allowed geneticists of the Forensic Genomic Consortium Netherlands (FGCN) to use sample/ID packages for scientific research.⁴⁵ As a result, molecular biologists currently conduct scientific research with sample/ID packages. These samples no longer refer to the donor's individual identity, yet the Custodian of the Dutch DNA database has provided information about the place of birth of the originators. This nominal information can be used to classify groups of people genetically. For example, if the FGCN research provides enough evidence that DNA profile 'MR' is typical for donors originating from the Moroccan Rif mountains, and DNA profile 'NA' is very common for donors originating from the Netherlands Antilles, then those two profiles provide information about external visible characteristics of people originating in those regions: an MR profile probably originates from someone who is fairly dark-skinned and has dark hair; an NA profile probably originates from someone who is black-skinned with black, curly hair. Such information could aid criminal investigation in general and the organization of a DNA dragnet in particular.

There is another issue at stake when sample/ID packages are genetically researched. Although the known body itself is not being touched when sample/ID packages are examined and hence – from a legal perspective – the inviolability of the person's body is no longer at stake, the body itself remains at issue. The reason for this is the synecdochal relation of the sample/ID package and the body where the former stands in for the latter.⁴⁶ When a sample/ID package is genetically examined, two civil rights are at stake. First, the privacy of the originator is at stake because his or her personal information contained in the genome can be revealed. And secondly, as result of the synecdochal relation between the known body and the sample/ID package, the *bodily integrity* – as opposed to the right to an inviolable body – of the originator arguably is violated when the sample/ID package is examined. The right to an inviolable body and bodily integrity are usually considered to be synonymous legal categories, yet this example

---

44 I. Van der Ploeg, 'Genetics, biometrics and the informatization of the body' (2007) 43 *Annali dell' Istituto Superiore di Sanità* 44–50; M'charek, op. cit., n. 24; Toom, op. cit., n. 27.

45 Dutch Ministry of Justice, 'DNA-onderzoek uiterlijk waarneembare kenmerken' ('DNA research external visible characteristics'), 9 February 2008, ref. no. 5528833/08; see, also, <www.forensicgenomics.nl>.

46 For a praxiographic investigation of synecdoche, see C. Cussins, 'Ontological Choreography: Agency through Objectification in Infertility Clinics' (1996) 26 *Social Studies of Science* 575–610.

shows that these categories can no longer be regarded as identical – forensic genetic bodies interfere with the constitutional categories themselves.[47]

Genetic research on reference samples originating from convicted offenders imports 'normative registers' different from those enacted and enabled by the known bodies discussed in the first two parts of this section.[48] First, research on reference samples violates the bodily integrity and privacy of the originators. Second, storing the sample/ID packages enacts another forensic genetic body that represents the diversity of the convicted offender population in the Netherlands, the sample/ID bank. Third, the originators of the samples are denied any kind of information, informed consent, autonomy, and control over their own cellular material, arguably violating dominant values in a biomedical context.[49] As a fourth kind of normative issue, it should be noted that knowledge about external visible characteristics does not stay within laboratory walls or the (electronic) pages of scientific journals. Instead, genetic markers for external visible characteristics will, in the Dutch context, be inscribed in the Law on External Visible Personal Characteristics. In addition, this knowledge will assist the organization of DNA dragnets. Every non-suspected individual who becomes associated with the suspect population because of race, sex or eye colour runs the risk of being included in a DNA dragnet. As these 'interesting' subjects do not fit any legal category, samples cannot be obtained by force, but should be volunteered. This brings us to a fifth normative issue: individuals who are associated to the suspect population are requested in criminal investigations to prove their non-involvement. This is a reversal of the onus of proof and implies an erosion of the presumption of innocence.

## CONCLUSION

Science, law, and biological bodies have been at the centre of analysis in this contribution. Instead of analysing them as separate realms, I have traced their interactions in practice and mapped their mutual effects. The analysis underscores science and law as productive forces giving shape to practices and their objects, in particular to the legal and genetic reconfiguration of bodies and bodily samples. I have analysed how the expansion of Dutch forensic DNA profiling has been dependent on access to bodies and bodily samples by focusing on three enacted forensic genetic bodies: the known

---

47 See, also, I. Van der Ploeg, 'Biometrics and the body as information: normative issues of the socio-technical coding of the body' in *Surveillance as Social Sorting. Privacy, Risk and Automated Discrimination*, ed. D. Lyon (2003) 67.
48 See Cole and Lynch, op. cit., n. 38; M'charek et al., op. cit., n. 39; Toom, op. cit., n. 27.
49 For an extensive comparison between forensic genetic and biomedical databases, see R. Tutton and M. Levitt, 'Health and wealth, law and order: banking DNA against disease and crime' in Hindmarsh and Prainsack, op. cit., n. 2.

body, a sample/ID package, and the sample/ID bank. These bodies are mastered by agents of power, are excluded from juridico-institutional models, and become known and 'public' by the production of genetic knowledge about them.

I have demonstrated different normative registers by contrasting the enacted forensic genetic bodies with entitlements to civil rights and legal principles. The right to an inviolable body and personal life are at stake when known bodies are enacted. This observation does not mean that forensic DNA typing is wrong – on the contrary. Yet, the empirical example of DNA hunts to collect known bodies was used to argue that there is a thin line between proportionate and disproportionate measures.[50] Sample/ID packages produce or convey another normative register. It was argued that synonymous constitutional categories such as the right to an inviolable body and bodily integrity can no longer be regarded as identical. When genetically examined, it is hard to maintain that the object of research is not the originator's body. This was exemplified by the research currently being conducted by the geneticists of the FGCN. The research is conducted without the originators knowing that their sample is used for scientific research, hence they are being denied information, informed consent, autonomy, and control over their own body samples. In addition, external visible characteristics are a technology enabling the organization of DNA dragnets, accompanied by a reversal of the onus of proof and weakening the presumption of innocence principle.

In conclusion, it can be stated that forensic DNA profiling has become an important aid to criminal investigation and litigation precisely because it centres on bodies and bodily samples in order to find the truth and provide material that can be used in the administration of justice. In that capacity, forensic genetic practices interfere with more conventional mechanisms deployed by the police in the course of criminal investigation, and have been dubbed 'genetic policing' or the 'genetic suspect'.[51] With the ongoing advancement of forensic DNA profiling in criminal justice systems, bodies and bodily samples will become ever more important markers for finding the truth and administering justice. As argued in this contribution, forensic genetic practices interfere with the distribution of civil rights and legal principles. As forensic science and enacted laws create new hierarchical relations between agents of power and private bodies, they are at the heart of democratic societies.

---

50 For the issue of proportionate measures in the context of English DNA profiling, see ECtHR, *Case of S. and Marper* v. *the United Kingdom* (application nos. 30562/04 and 30566/04) 4 December 2008.
51 R. Williams, 'DNA databases and the forensic imaginary' in Hindmarsh and Prainsack, op. cit., n. 2; Williams and Johnson, op. cit., n. 2; Hindmarsh and Prainsack, op. cit., n. 2.

## The Materiality of What?

ALAIN POTTAGE*

*A singularly influential sense of 'material worlds' has been developed by actor-network theories of science and technology, which trace out the kind of social action that emerges from encounters between 'humans' and 'non-humans'. What happens when this approach to materiality takes on the question of law? One answer is suggested by Bruno Latour's recent ethnography of law making in France's Conseil d'Etat. Interestingly, this study turns out to be not so much an actor-network theory of law as occasion to add a new dimension to the material worlds of actor-networks, namely, the communicative dimension of 'regimes of enunciation'. My hypothesis is that this distinction between the sociality of actor-networks and the logic of enunciation is problematic because it uncritically adopts the premise that there is an institution such as 'law' that has to be explained or materialized by social science, thereby diminishing the critical energy that the theory of actor-networks or of* dispositifs *might bring to the study of law.*

For some time now, social studies of science have explored the involvement of material things in the fabrication and reproduction of scientific knowledge. This material agency has been construed or schematized in different ways, and materialities have gone by different names, but the broad effect has been to develop an extraordinarily productive critique of the assumption that scientific or technological knowledge is a product of human agency or intentionality alone, or an effect of compromise between purely human institutions or collectivities. Bruno Latour's choreographies of human and non-humans, or of hybrid actants, now pitched as an 'ecological' or 'compositionist'[1] politics, have popularized the idea that any mode of

* Law Department, London School of Economics, Houghton Street, London WC2A 2AE, England
r.a.pottage@lse.ac.uk

1 B. Latour, 'An attempt at a compositionist manifesto' (2010) 41 *New Literary History* 471–90.

analysis has to include the agency of artefacts in its dramatization of the social. This sense of the materiality of 'material worlds' suggests one possible script for a conversation between science studies and 'law studies'. The fabrication of legal knowledge also involves materialities of various kinds[2] – space, bodily *hexis*, archives, databases or archaic loose-leaf binders,[3] forensic models, files, sketches, inscriptions – which have become the focus of a number of studies. Two notable examples are Thomas Scheffer's microsociology of the criminal trial and Cornelia Vismann's historical studies of forensic and bureaucratic media.[4] So the central question in this conversation might be whether law has any technologies at all, in the sense of material agencies that inflect or 'shift' human action, or whether, as Latour suggests, law has only discursive materialities of various kinds, and, if so, what the agency of these scriptural or semantic media actually is. Is law a 'material world' in the same sense as science or technology?

First, what do we mean by 'materiality'? Most studies of material agency in science begin in the midst of things, with a close ethnography or detailed text-based reconstruction of the density, conformation, disposition, and operability of technical media or devices, and of the gestures, perspectival axes, and textual traces that unfold around these material agents or 'non-humans'. The point is to begin with the actors themselves, these being such things as spectrometers, micropipettes, electrophoresis gels, or neuronal tissues, as much as human agents or intellects. But what really matters is not the simple materiality of these things – their mass, density, or spatial definition – but rather 'materiality' as the kind of agency that is afforded by, elicited from, or ascribed to them. Indeed, material agency is not an innate quality of these artefacts. Bruno Latour's notion of hybrid actants makes it clear that agency is not inherently either human or non-human; it is an emergent effect of the composition of humans and non-humans, or of their reciprocal engagement or co-variation as moments in the unfolding of an actor-network. Things or artefacts – scientific or technical materialities – point beyond themselves to contingent processes of 'sociality'. In other words, 'materialities', as the points in which the transition of these processes become visible and traceable, become ciphers for 'materiality' as a kind of

---

2 For some studies of these legal or forensic materialities, see, for example, A. Pottage, 'Law machines: scale models, forensic materiality, and the making of modern patent law' (2011) 41 *Social Studies of Science* 621–43; A. Riles, 'Introduction: In response' in *Documents: Artifacts of Modern Knowledge*, ed. A. Riles (2006) 1–38; T. Scheffer, 'Materialities of legal proceedings' (2004) 17 *International J. for the Semiotics of Law* 365–89; C. Vismann, *Files. Law and Media Technology* (2008); E. Weizman, 'Forensic Architecture: Only the Criminal Can Solve the Crime' (2010, Nov/Dec) *Radical Philosophy* 9–24.
3 See H.T. Senzel, 'Looseleafing the Flow: An Anecdotal History of One Technology for Updating' (2000) 44 *Am. J. of Legal History* 115–97.
4 See, respectively, T. Scheffer, *Adversarial Case-Making. An ethnography of the English Crown Court* (2011) and C. Vismann, *Medien der rechtsprechung* (2011).

dynamic condition of existence. Hence the proposition that 'materiality is sociality',[5] or that 'to rematerialize is to resocialize, to resocialize is to rematerialize'.[6] Ultimately, 'materiality' becomes a signifier of contingency, of *'ce qui fait que tout se fait'*.[7]

Perhaps because law's materialities – essentially, enunciations articulated into a limited range of media – are relatively unglamorous, studies have evolved a somewhat more plural, exploratory, and open-ended approach to the question of material agency. For example, Thomas Scheffer's study of the role of inscriptions and narratives in the preparatory phases of a criminal trial mobilizes a broad range of theoretical idioms or perspectives to trace out the complexity of material agencies. Michel Foucault's figure of the *énoncé*, as a discursive form which is material because it affords certain possibilities of 'reinscription and transcription',[8] captures the material agency of stories or narratives as emergent framings of encounters in the trial process, Gilles Deleuze's theme of 'becoming' informs the apprehension of files or texts as 'relative becomings', or as forms that are materialized temporally, and Niklas Luhmann's theory of 'materialities' as structural redundancies or as correlates of observation[9] functions as a kind of 'control' on other senses of materiality. This kind of approach – which invokes no big signifiers and offers no ecumenical politics – has the virtue of recognizing materiality as a technical, semiotic, mediatic, phenomenological, cybernetic, and temporal complex.[10] Taken in this way, materiality is a theme with considerable critical potential, which gives us the resources to dissolve and recompose the premises or taken-for-granted categories that intervene before analysis gets under way.

This is precisely what Latour's actor-network theory achieved in the context of science studies, by introducing artefacts – non-humans – into the

---

5 B. Latour and V. Lépinay, *L'économie, science des intérêts passionnés* (2008) 47.
6 B. Latour, 'Les baleines et la forêt amazonienne. Gabriel Tarde et la cosmopolitique' (interview with Erin Manning and Brian Massumi) (2009) 3 *Inflexions* 1–16, at 10.
7 H. Bergson, cited in M. Lazzarato, 'Pour redéfinition du concept de biopolitique' (1997), at <http://multitudes.samizdat.net/Pour-une-redefinition-du-concept>.
8 Scheffer, op. cit., n. 2, especially at p. 380:
   [A story might be] employed several times during the pre-trial: in the police interview, the primary disclosure, the defence statement, the brief to counsel, the barrister's notes, plea bargaining, etc. Every employment triggers the story's re-appearance and modification, and hence, continuation and imposition.
   See, also, generally, M. Foucault, *L'archéologie du savoir* (1972).
9 See, generally, N. Luhmann, *Social Systems* (1995). For systems theorists, materialities have to include their own observation; materiality only exists as a referent of communications, which necessarily presuppose an observer or an idiom of observation.
10 One might say the same of the approach taken by Cornelia Vismann, whose work has obvious affinities with Friedrich Kittler's historical analysis of communications and, in particular, the medial schemata that constitute the 'historical' or 'anthropological a priori' for the existence of 'so-called Man' (see F. Kittler, *Gramophone, Film, Typewriter* (1990) 29 and 117, respectively), but also incorporates perspectives in everything from psychoanalytical theory to actor-network theory.

understanding of science and technology, and by introducing the energy of Deleuzian assemblages into a largely instrumentalist representation of artefacts and their politics. But, instead of folding law into the dynamism of actor-network sociality, Latour takes it as the occasion to unveil a new dimension of his compositionist politics, namely, the dimension of regimes of enunciation or modes of existence. Regimes of enunciation precipitate from the sociality of actor-networks, almost as second-order commentaries on this 'original' mode of sociality, and they unfold according to the principles of discursive engineering that hold together such things as statements, speech acts, and shifters. Although law as a regime of enunciation is supposed to emerge from the sociality of actor-networks, it is not clear by what means a regime of enunciation that construes itself as autonomous actually (re)engages with what systems theorist would call its 'environment'. Indeed, Latour's analysis of law often proceeds as though there were actually nothing more to law than a process of enunciation. The effect is to suggest that law is not a material world in the same sense as science or technology; or, perhaps, that the study of law reveals a dimension of society that should now be reintroduced into our apprehension of technoscientific sociality. My argument is that this representation of law as a regime of enunciation is too indulgent of the lawyer's sense of law, and gives too narrow a sense of the rhizomatic *dispositifs* in which legal forms or materiality are implicated. Instead of seeking to materialize or substantiate 'law' as a kind of universal category, why not mobilize materialities to develop alternative and more plausible ways of tracing out these implications?

## LAW AS ENUNCIATION

Latour observes of the Conseil d'Etat that 'he [the ethnographer] had never participated in an institution that had so little concern at being studied, or that was so indifferent to external observation'.[11] But the indifference of the institution turned out to have a methodological virtue, which was to reveal the Conseil as 'an ideal way into the legal mode of veridiction'.[12] The ethnographer could then go on to distinguish law as a regime of enunciation or veridiction from the institution in which it was articulated:

> I extracted the work of law from the institution in much the same way as a physiologist might have extracted the spinal cord of a dog, knowing perfectly well that it was not the whole animal.[13]

So, whereas the study of, for example, the Aramis project took Latour into design laboratories, manufacturing workshops, and the offices of politicians

---

11 B. Latour, *La fabrique du droit: Une ethnologie du Conseil d'Etat* (2002) 268.
12 id., p. 271.
13 id.

and administrators,[14] the study of law in the Conseil was focused exclusively on transactions within the chambers of the Palais-Royal. Law making was reduced to what could be observed from the margins of this 'cold' institution; namely, a set of transactions in gesture, speech and text, the material processes of compiling and circulating files, and some effects of architecture and posture. On this basis Latour suggests that instead of imagining law as a corpus of rules that actually has the capacity to bind people or events in the world, we should construe law as the art of binding – or concatenating – statements or communications:

> The set of functions that permit one to relay, retrace, hold together, attach, suture, or stitch back together what it is in the very nature of enunciation to separate or distinguish, belong to the technique of attachment which our western tradition celebrates as law.[15]

How does the classic actor-network sense of materiality-sociality figure in this analysis of law as a technique of enunciation?

One answer is that the study of law makes the same turn away from the 'society' of sociologists as did the study of science:

> if the study of science and technology compelled us to abandon the sociology of the social in favor of the sociology of association, then the analysis of law encourages us even further in that direction.[16]

One of the objects of Latour's studies of science was to get away from the assumption of 'society', with its configurations of structures, forces, fields, or systems, as a pre-existing frame or landscape for social action, and to open up instead the idea of a sociality of association, in which networks are an emergent product of the association between actors whose competences are themselves emergent effects of association. If 'society' is 'the reciprocal possession of one by many, and many by one [*'la possession réciproque, sous des formes extrêmement variées, de tous par chacun'*]',[17] then it is not a 'whole' with pre-ordered 'parts' but a multiplicity in which the 'whole' circulates only as the diffracted set of representations that each monadic individual has of it. In any case, the sociality of legal enunciation is not quite like the sociality of science; it is        not just another variation on the same modes of 'associative' materialization. In science, association is an effect of relations of irritation, inflection, surprise, or recalcitrance that emerge from the 'technological' engagements of humans and non-humans, and, according to Latour, law has no technologies of this kind:

---

14 See, generally, B. Latour, *Aramis, or the love of technology* (1996).
15 B. Latour, 'Note brève sur l'écologie du droit saisie comme énonciation' in *Pratiques cosmopolitiques du droit*, eds. F. Audren and L. de Sutter (2005) 34–40, at 34.
16 Latour, op. cit., n. 11, p. 280.
17 See B. Latour, 'La possession. Une preuve par orchestre' in *Anthologies de la possession*, ed. D. Debaise (forthcoming).

> [E]ven the most humble technology – a lamp, an ashtray, a paperclip – mingles periods, places, entirely heterogeneous materials, folds them into a single black box, and prompts those who use it to act by inflecting their course of action. Law is not capable of doing this. It is the least technological [*technique*] of all forms of enunciation.[18]

One might counter that law is a technology in the sense that ancient rhetoric was a *techne* – an art of producing things that could as well be as not be[19] – but, for Latour, this would a *techne* of the time-space of text rather than the time-space of material worlds.[20] Law's rhetorical or enunciative mode of articulation does not have the most basic ingredients of 'association' as revealed by the study of science.

In fact, Latour's study of law inaugurates a second phase of what has come to be presented as an 'ecological' or 'compositionist' politics. The question for this politics is 'first, to define the beings that we have to assemble so as to make them compatible with one another, and, second, to distinguish the different ways in which they are assembled'.[21] The beings in question are defined by the sociality of actor-networks, as the association of actants, monads, humans or non-humans that is revealed by the study of technoscientific networks, but their modes of assembly involve a very different mode of agency or association, one that ultimately has to do with the of linguistic or discursive media. Latour adopts an appropriately materialist metaphor to characterize this difference:

> We can agree that institutions such as Science, Religion, and Law are mingled indistinctly, rather like the veined marble panels of the Basilica di San Marco in Venice, in which no figure is clearly recognizable (indeed, the intuitions of actor-network theory were first prompted by this kind of commingling). But this does not answer the question of their truth criteria and their respective felicity conditions, because one particular regime always plays the role of a

---

18 Latour, op. cit., n. 11, p. 293.
19 See R. Barthes, 'L'ancienne rhétorique' (1970) 16 *Communications* 172–223.
20 B. Latour, 'Where are the missing masses? The sociology of a few mundane artifacts' in *Shaping Technology/Building Society: Studies in Sociotechnical Change*, eds. W.E. Bijker and J. Law (1992) 225–58, at 239:
> Engineers constantly shift out characters in other spaces and other times, devise positions for human and nonhuman users, break down competences that they then redistribute to many different actors, and build complicated narrative programs and subprograms that are evaluated and judged by their ability to stave off antiprograms. . . . Instead of sending the listener of a story into a different world, the technical shifting-out inscribes the words into another matter.

It might be, however, that the space-time of text is the basic archetype; Latour's studies of the scientific laboratory were primarily about inscriptions, about differences and tensions between texts and inscriptions, or about the 'cult of the inscription' in the laboratory (see H. Schmigden, 'Die Materialität der Dinge? B. Latour und die Wissenschaftsgeschichte' in *Bruno Latours Kollective: Kontroversen zur Entgrenzung des Sozialien*, eds. G. Kneer, M. Schroer, and E. Shüttpletz (2008) 15–46).
21 Latour, op. cit., n. 15, p. 34.

dominant [in the musical sense], which is why I say that in the Conseil d'Etat the decision between what is true and what is false is made legally, in a way that is obviously not religious, scientific, technological or political.[22]

The original sociality of the actor-network is essentially unpatterned; institutions such as law or science exist, but their configuration and extension are qualities that depend on how they are traced out, either by those involved in these institutions or by more panoptic actor-network theorists. Modes of existence emerge from within this marbled sociality as techniques of enunciation that organize the 'rest' of sociality into patterns that are made according to the particular 'truth criteria' or modes of attention that are articulated by each. Institutions run into each other confusedly; modes of enunciation stand out in clear relief. Law as it emerges from the study of the Conseil is just such a 'mode of existence', or 'mode of enunciation' and, like other modes of existence, it comes into being by detaching or differentiating itself from within the 'sociality' of assemblages, hybridizing inflections, delegations, and human/non-human associations.

Law is in every sense the most exemplary 'mode of enunciation'. At least since the advent of legal positivism, law has been cast as an institution, regime, or system that exists only as an effect of self-description; that is, law comes into being paradoxically, as an effect of the identification of certain enunciations or transactions as 'legal' by reference to a criterion that is posited by those enunciations or transactions themselves. And Latour presumes precisely this effect in framing his ethnographic observation of action in the Conseil d'Etat; the identification of certain communicative, gestural, or material transactions as the elements of a 'legal' regime of enunciation depended on ability of the ethnographer to rejoin, and in some sense to reaffirm, the internal point of view of these communications, which all invoked 'Law' as their ultimate addressee.[23] Already, then, law is a communicative or enunciative artefact, and one implication of its constitution through self-description is that it precipitates from the material worlds of actor-network sociality by translating, reconstructing, or re-engineering the sociality of actor-networks into the dynamics of communicative or enunciative action.[24] In one sense, this affirms Latour's point that just as the fabrication of science cannot be explained by reference to 'society', nor can the fabrication of law; after all, 'there is more "society" in law than there is in the society that is supposed to explain the making and operation of law'.[25]

---

22 id., pp. 34–5.
23 For a different ethnographic construction of an aspect of lawmaking, see A. Riles, *Collateral Knowledge. Legal Reasoning in the Global Financial Markets* (2011).
24 Hence the observation that, rather than resolving disputes, legal procedures actually 'expropriate' the conflict and its parties by parasitizing their 'energy' (see G. Teubner, 'Alienating Justice: On the surplus value of the twelfth camel' in *Law's New Boundaries: The consequences of legal autopoiesis*, eds. J. Pribán and D. Nelken (2001) 21–44.
25 Latour, op. cit., n. 11, p. 278, referring to Yan Thomas's studies of Roman law.

But the analogy between science and law obscures a quite radical change in the theoretical agenda. Law makes the point so compellingly because autonomy is implicitly or explicitly asserted in everything that lawyers themselves say when they perform the work of law. And, by contrast with science, to know what law is, one has to join this 'internal point of view' rather than simply observe what lawyers do. The 'society' that exists in or for law – that is to say, the semantic or phenomenal content of law's society – is an effect of processes of materialization, inflection, or detour (to borrow some Latourian terms) that are very different from those of the techno-scientific actor-network. We switch from the space-time of materialities in the sense of actor-network theory into the space-time of speech acts, statements, or *énoncés*, and hence from one kind of medium to another.[26] So, if law is taken to be a distinct mode of enunciation, how is enunciation articulated; or, more precisely, what is its materiality?

## MATERIAL WORDS

The continuity or connectivity of law as a regime of enunciation presupposes the existence of what Latour calls a code – *une clef de lecture* – which indicates that one is dealing with a legal enunciation:

> Here, I claim to have given a possible explanation for the tautology that is common to all definitions of law, a tautology that strikes both specialists and outsiders; the tautology arises because one cannot understand any particular act of the institution of law unless one adds the following code [*clef de lecture*]: 'what you are about to read or hear is Law rather than fiction, or politics, etc …'.[27]

Law is defined not by the propositional content of enunciations but by a marker that qualifies a given enunciation as an element of law and not of some other mode of existence (such as politics, economics, or religion).[28] The distinction between law as an institution and law as a mode of enunciation is premised on this distinction between content and code. Codes of enunciation structure the contingency of communicative processes; precisely because it is an operation of transmission, enunciation necessarily creates a split between utterance (the act of enunciation) and information (the content of enunciations). And, if 'it is in the nature of enunciation to send or transmit, and so to break the connection between speaker and what is said',[29] then the (re)connecting of information to utterance so as to create

---

26 Presumably, law might also exist as an institution, and hence as an actor-network, but Latour's representation of law as a mode of enunciation, or as a mode of binding or concatenating statements, makes it difficult to see how one would retrace one's steps back to this other existence.
27 Latour, op. cit., n. 15, p. 36.
28 For the case of politics, see Latour, op. cit., n. 6, p. 10.
29 Latour, op. cit., n. 15, p. 37.

meaning is necessarily an effect of how that (re)connection is made, and, by implication, meaning is contingent on the observer or addressee who makes the connection. Statements have to be ascribed or imputed to speakers, and ascription is itself a selective or contingent operation. So how is this process of contingent selection – the art of connecting enunciations – figured in Latour's reconstruction of law? How, precisely, does a *clef de lecture* function to differentiate legal enunciations from political, economic, or religious modes of existence?

Latour mobilizes another materialist metaphor to explain what distinguishes 'law' from other regimes of enunciation (in this case, politics):

> Imagine a game of LEGO in which the traditional attachment by means of four studs is replaced by attachments of many different kinds. Imagine then that each of these attachments makes further attachments either easier or more difficult. Now assume that in this somewhat peculiar game of LEGO some blocks are connected by means of a LAW connector and others by means of a POL connector. The blocks themselves are of different shapes. Give the game to some kids to play with. They will produce forms – institutions – which will have longer or shorter segments which we can call LAW because they are connected by means of a LAW attachment, even though a given block might also, in another segment, be joined by means of a POL attachment. Of the multicolored assemblage that is produced, one might say, depending on the intensity of the connections, 'that, more or less, is law', and 'that, more or less, is politics'. This will never be entirely true, because the blocks will be of different shapes and colors, but at the same time it will not be entirely wrong, because the 'dominant', to adapt a musical term, will indeed be given by a particular kind of attachment, perturbation, or contamination.[30]

Although the last words of this excerpt – 'perturbation, or contamination' – evoke the contingent sociality of actor-network theory, the metaphor materializes the contingencies of enunciation at precisely the point at which one might expect a theoretical account of how that contingent sociality plays out in the medium of communication or enunciation.

The difficulties of this materialist version can be brought out by comparing Latour's take on enunciation with Niklas Luhmann's theory of the communicative media and transactions that maintain social systems. Latour's compositionist politics is constitutionally allergic to Luhmann's theory of autopoietic systems, perhaps because of the abstract meta-language in which it is presented, and perhaps because materialities, or the actors themselves, are so decisively eclipsed by the existence that they come to have in the schematic horizon of the observer. Although much of what Latour has to say about the existence of law in society is entirely consistent with Luhmann's theory of law as a social system,[31] Latour rejects the

---

30 id., p. 40.
31 See, generally, N. Luhmann, *Law as a Social System* (2007) and G. Teubner, 'Global Bukowina: Legal pluralism in the world society' in *Global Law Without a State*, ed. G. Teubner (1997) 3–28.

systems-theoretical perspective on law.[32] Confusing Luhmann's social 'systems' with Pierre Bourdieu's 'fields',[33] Latour observes that the notion of differentiated social systems is even less plausible in the case of law than in the case of science because in the case of law one does not have 'the pretext of the clear line of demarcation made by walls of the laboratory and the white coats of laboratory workers'.[34] The conclusion is that Luhmann's theory overstates the premise of autonomy: '[L]aw is autonomous in relation to social because it is a means of producing the social, of articulating and contextualizing it, [but] it has no specific domain or territory'.[35] But Luhmann's theory of law – and of social systems more generally – says almost exactly that, and, more importantly, it introduces into the explanation of law as a communicative system or 'regime of enunciation' precisely the kind of contingent or emergent action that is found in actor-networks and that is suspended by the materialist metaphor of communication as a construction of LEGO blocks.

For Luhmann, to the extent that law exists, it exists virtually, as 'know-how' that is actualized, or communicated into existence, only when and for so long as it is used as a code or medium of communication. The material and other premises of law – archives, standardized documents, semantic forms – take effect as 'law' through this process of actualization. Contrary to what Latour suggests, law is not construed as a quasi-spatial territory or a bounded domain. Systems theory is not about drawing boundaries around systemic territories; it is about making distinctions, and distinctions are not boundaries but differences made by observers, each of whom schematizes the space and time within which 'boundaries' are drawn. Second, to trace out the functioning of law, one needs to observe the reciprocal modes of coupling and co-production by which the codes, media, and materialities of law are implicated in a broader sociality, and which condition the communication of 'law'.

Luhmann's systems theory resolves the process of communication into three contingent terms or selections: information, utterance, and understanding. The values of each of these terms emerge from their articulation in communication as a mode of double contingency – crudely, the process of 'seeing oneself being seen', to use a formula that is somewhat too Hegelian. From the perspective of the addressee, the 'meaning' of an enunciation in Latour's sense is an effect of how the two selections of utterance and information are spliced together: as speech act theory tells us, the informational content of a statement depends on who utters it, and how. From the

---

32 For a more sophisticated exploration of possible convergences between actor-network theory and systems theory, see G. Teubner, 'Rights of Non-Humans? Electronic Agents and Animals as New Actors in Politics and Law' (2006) 33 *J. of Law and Society* 497–521.
33 Latour, op. cit., n. 11, pp 282–3, fn. 47.
34 id., p. 283.
35 id.

perspective of the speaker, the question is how to anticipate how the addressee will splice utterance and information together, and to how to modulate those terms accordingly. Luhmann's formula of double contingency has some resonances with Latour's sketch of the disruptive effects of enunciation, but it is the premise of a very different, and much more complex, account of how contingencies of communication or enunciation are patterned into differentiated regimes such as 'law'. To focus on the central figure in Latour's account of law, the premise of double contingency leads to a very different sense of how communication is 'coded'.

From the perspective of systems theory, Latour's characterization of the *clef de lecture* that marks legal enunciations is too one-dimensional. It misses a crucial point about the practical operation of law. The legal quality of enunciations is not given in advance; it is always an effect of ascription. In law, the juridicality or non-juridicality of any event is always in question, so the first and most basic technique of law is to decide on the difference between legal and non-legal. The basic technique of law is not to connect ready-made blocks of legal enunciation into chains but to produce legal enunciations by qualifying events or enunciations as legal in the first place. Law makes its own 'building blocks' and connections between blocks. Events or enunciations that have not so far been construed as legal might suddenly, depending on how the distinction is drawn, be qualified as legal: one might find that one has made a contract by acts that had not so far been taken to manifest the requisite intention; the 'facts' of a case might be fictionalized so as to ascribe legal effect to a 'non-juridical' act or event.

To account for this basic feature of law, Latour's *clef de lecture* would have to be reconstructed as a binary code rather than univocal marker. The code that identifies legal enunciations is not attached to enunciations in the quasi-material sense suggested by Latour's Lego metaphor; it is only one term of a distinction that produces 'law' by distinguishing it from what is qualified as 'non-law'. Given that distinctions are not inscribed in the world, but are inflicted upon it by observers, law exists only as the competence or know-how that animates its founding distinction. And legal know-how is not an attribute of human actors; it is immanent in a medium which prefigures any of its users, and which has vastly more complexity or potentiality than any participant could possibly grasp.[36] So law's *clef de lecture* cannot exist as a quasi-material form, but only as a disembodied, dematerialized, and 'non-human' structure that is sustained by the recursive operations of a system which, in the old adage of systems theorists, 'produces itself from its own operations'.

Presumably, Latour would agree that *clefs de lecture* are not inherent in propositions but have to be ascribed to them. Propositions have to be

---

36 On the shift from materiality to media, see J. Clam, *Trajectoires de l'immatériel. Contribution à une théorie de la valeur et de sa dématérialisation* (2004).

'coded'. But if codes are binary rather than singular, and if the process of coding presupposes the observer who distinguishes law from non-law and the medial competences mobilized by that observer, then we need to draw on something other than Latour's sense of sociality as inflection to characterize the process of coding. The associative logic of networks has to be transformed into, or complemented by, an account of the communicative logic of enunciation.

All of this will be familiar to theorists of law. The point of the contrast between 'modes of existence' and 'systems' is not to argue for one or other of these reconstructions of law, but to point out that Luhmann's theory of law offers the most astute and consistent answers to the questions that are latent in Latour's sketch of law as a mode of enunciation. Or, to put it the other way around, we can see in Luhmann's systems theory the kind of theoretical moves that actor-network theory would have to make if it were to develop a theory of law as the paradigm of a mode of enunciation. Quite simply, systems theory proposes the most plausible realization of Latour's proposition that 'law juridifies all of society, which it apprehends as a totality in its own particular way'.[37] If we are to materialize law as a communicative system or regime of enunciation, then systems theory offers the most plausible account of materiality as momentum; or, more precisely, as an effect of 'time-binding', in which each communication binds time by defining the horizon against which the succeeding communication proceeds.[38] But, keeping in mind the theme of 'material worlds', the real question is whether we should mobilize the rich conceptual potential of theoretical reflections on materiality simply to give substance to the assumption that there such a thing as 'law'. Why not instead recruit the potentiality of 'materiality' to imagine material worlds that are not always already configured into law, science, politics, and so on, but which are, in something like the original sense of actor-network theory, as confused as the veining of the marbles of St Mark's Basilica, and which call for more productive and more adequate modes of analysis?

## MATERIALIZING LAW

Why is the premise or assumption of 'law' so problematic? For decades, if not longer, the purpose of 'critical legal studies' or 'socio-legal studies' of diverse persuasions has been to tell us something more or different about the

---

37 Latour, op. cit., n. 11, p. 281.
38 One might ask, from the perspective of Kittler's analysis of media technologies, whether systems theory can really observe 'the communication of communications': see G. Winthrop-Young, 'Silicon sociology, or, two kings on Hegel's throne? Kittler, Luhmann, and the posthuman merger of German media theory' (2000) 13 *Yale J. of Criticism* 391–420.

social phenomenon of 'law', and these diverse lines of inquiry all presume one thing – the existence of 'law' as an instance or phenomenon whose existence is too evident to require justification. The effect of successive theoretical, sociological, and anthropological investigations and critiques has been to expand law beyond the classic scenes of legislation and adjudication, 'beyond the state', and into the texture of a social life that is thereby anatomized (*sub silentio*) in such a way as to make it the medium (or, more accurately, the 'context') of this expanding and ever-ramifying instance. Law has been traced into the plane of the unconscious, materialized in architectural forms, cast as the expression of 'cultures' of diverse or plural kinds, constituted as the correlate of some negated 'other', and generally replicated into the social by means of theoretical devices such as analogy or change of scale, etymology or genealogy, and various modes of differentiation and coupling. This process of abstracting laws into 'law', or of absorbing legal forms into their animating contexts, has been taken as the hallmark of progressive or critical thought. These moves have turned 'law' into an abstract and generalized social instance, or a question that exists even before theoretical reflection gets under way. Both Latour and Luhmann radicalize the question of law in ways that call into question the old 'socio-legal' or 'law and society' agenda; for both, there is indeed more society in law than there is in the society that is invoked by studies in 'law and society'. Ironically, however, the effect is actually a retrenchment of the premise that law exists as a singular social instance that it is the business of the theorist to explain.[39]

So we have two contrasted strategies for realizing law as a social instance. One is based on Luhmann's sense of what it is for a theory to be well made; a theory has to be adequate to the complexity of contemporary society, to account for the paradoxical and emergent nature of social beings and social processes, and have an architecture that is thoroughly self-reflexive. The other is informed by Latour's sense of what it means for the ingredients of sociality to be 'well composed', or 'put together while retaining their heterogeneity',[40] which implies a somewhat different sense of reflexivity. Any theory of (say) law or science should be able to engage the attention of the actors themselves, and to do so in such a way as to encourage them to perceive the institution in which they are involved as one of many 'modes of existence', and hence to perceive themselves as participants in a world that has to be 'composed' in the style of Latour's ecological politics. These very different takes on materiality converge almost inevitably on the instance of 'law'. What happens if we reverse the direction of this movement, and instead of presuming 'law', and asking how materiality should be configured

---

39 Latour (op. cit., n. 11, p. 282) observes that Luhmann starts from 'the most reductive and often the most ethereal definition that the domain of law applies to itself' but, in a sense, the same is true of his own theory of law as a mode of enunciation.
40 Latour, op. cit., n. 1.

to make good on that presumption, why not begin with the extensive potentialities of 'materiality' and ask what becomes of 'law' if we try to hold those potentialities open? Why strain so hard to materialize law?

Precisely because it proposes such an astute theory of contemporary law, systems theory also offers the best illustration of the retrenchment of 'law' as a conceptual premise. Systems theory has explored modes of 'law-making' that are considerably more varied and complex than the mode of enunciation that Latour draws out of the Conseil d'Etat. In particular, it has paid close attention to configurations in which one can see the 'polycontextual' logic of law in society. The central example is transnational law, which emerges from the association of thoroughly heterogeneous actors: states, corporations, NGOs, international organizations, indigenous peoples, and so on. These actors are not organized by reference to the warrants of the old juridical schema of sovereignty as hierarchy or center-periphery: territory, constitution, state. Rather, as in an actor-network, the competences and attributes are not ordered by any presiding instance but emerge (paradoxically) from the ways that particular transnational regimes organize themselves. And 'law-making' of this kind is necessarily co-produced with other kinds of social transaction: economic, political, scientific, and so on.[41] From this perspective, transnational legal ordering reveals a truth about law that is mystified by the self-representation of institutions such as the Conseil d'Etat. Yet, in dealing with the complexity of this mode of legal ordering, systems theory artfully reconstructs law from the diffracted elements of transnational law; law resurges wherever actors employ the distinction between legal and non-legal (enunciations). However complex any given articulation of law in society might be, the theorist will be able to peel away a diagram tracing out the specifically legal dimension of transactions within this articulation and hold that diagram up to anyone who claims that transnational regimes are just about economic power or 'social norms'. Ultimately, what is really 'radical' about this theory is its capacity to rediscover or reconstruct law in circumstances in which it might seem to have disappeared.

What if one were to reverse the trajectory of the inquiry, and instead of asking how the elements of the conceptual assemblage of 'materiality' should be mobilized to realize law as a social instance, ask whether a reflection on materiality might not actually lead to the dissolution of law as a social instance? At the risk of being taken as a reactionary Latourian, this is precisely what I should like to propose. My starting point is the old kinship that Latourian actor-network theory had with Foucault's conception of the *dispositif*.[42] The fullest definition of a *dispositif* as an apparatus, assemblage, arrangement, network, or device appears in an interview given by Foucault in 1977, shortly after the publication of the *History of Sexuality*:

---

41 The classic text is Teubner, 'Global Bukowina' (op. cit., n. 31).
42 For a brief overview, see J.-S. Beuscart and A. Peerbaye, 'Histoires de dispositifs' in (2006) 2 *Terrains & Travaux* 3–15.

[F]irst, an essentially heterogeneous ensemble, composed of discourses, institutions, architectural formations, regulatory decisions, laws, administrative measures, scientific statements, philosophical, moral, and philanthropic arguments; these are the elements of a *dispositif* – in short, what is said as much as what is unsaid [*du dit aussi bien que du non-dit*]. The *dispositif* itself is the network that might be established between these elements. Second, what I want to identify in a *dispositif* is precisely the nature of the linkage that exists between these heterogeneous elements ... These elements, whether they are discursive or non-discursive, are linked by something like a game, with changes of position or modifications of functions that can themselves be very different. Third, by *dispositif* I mean a kind of formation which has in a particular historical moment been given the important function of addressing some kind of some urgent situation. Therefore, a *dispositif* has a predominantly strategic function, [which involves] a rational and concerted intervention in relations of force, either so as to develop them in a particular direction or so as to block them, stabilize them, or exploit them.[43]

As in the case of Latour's actor-networks, *dispositifs* are 'assemblages [that] are made up of nothing other that what they assemble';[44] strategic or tactical modes of actualizing and conjoining the elements of a *dispositif* are emergent articulations that are conditioned by the very elements that they purport to organize.

What becomes of 'law' in the configuration of a *dispositif*-network? First, instead of presuming 'law', one would begin with a set of raw elements: texts, institutions, statements, gestures, architectural and material forms, formalized roles and competences, and self-descriptions (people often characterize themselves as practitioners or participants in 'law'). And, instead of abstracting to a field, medium, code or rationality in which these elements cohere into 'law', one would explore the ways in which elements are assembled into *dispositifs*. Most theories of law focus on documents (or their analogues) and the things that lawyers do with or around documents and texts: drafting, negotiation, research, formalization, interpretation, archiving, citation, argumentation, and so on. Indeed, the question of materiality is interesting to socio-legal scholars precisely because it suggests ways of reconstructing these basic operations. What is questionable is whether we need to continue integrating these operations into law as an instance, institution, or 'mode of existence'. The reference to Foucault's *dispositifs* is apposite because their object was to 'replace what Foucault defined in critical terms as universals'; namely, 'general categories or rational entities such as the state, sovereignty, law, or power'.[45] If we think of the most celebrated of Foucault's *dispositifs* – discipline, sexuality, governmentality – then each of these incorporated or metabolized law, but they did so precisely by treating texts, practices, visibilities, and self-descriptions as elements that derived their sense and effect from their articulation in each given *dispositif*. The

---

43 M. Foucault, 'Le jeu de Michel Foucault' in *Dits et Ecrits*, Vol. iii (1994) 299.
44 B. Latour, *We Have Never Been Modern* (1993) 138.
45 G. Agamben, *What is an Apparatus?* (2009) 23.

elements that we conventionally (or radically) materialize as 'law' were instead materialized in *dispositifs* that were inventive and yet recognizable.

Systems theorists will point out that *dispositifs* are arbitrary because their makers can give no account of how they come into being or are stabilized, other than by invoking some immanent principle of 'power', 'strategy', or, perhaps, 'materiality'. Those who share Latour's view that theory should instead be reflexive in the mode of 'diplomacy' – one has to engage the actors themselves rather losing them in a rarified meta-language – will say that the figure of the *dispositif* is too complex. And, given the somewhat disappointing results of the generalization of the notion of the 'assemblage' in the social sciences, one might ask whether the resurrection of Foucault's *dispositifs* would really be so productive. The answer is that it depends on how these *dispositifs* are made. It is true that the figure is somewhat indeterminate; as Giorgio Agamben remarks, any one *dispositif* can be resolved into many, depending on the order of the analysis or the interests of the observer, and the direction of scaling can be reversed. It is also true that there is no ready-made theoretical formula for a good *dispositif*; Le Corbusier once observed that buildings were not things that one talked about, but things that one walked through (*'on ne discourt pas sur un bâtiment, on le parcourt'*). One might say something similar of Foucault's *dispositifs*. They are artefacts whose conceptuality is expressed in the mode and effects of their assembly, and the singular art of Foucault was to invent figures that were both radically new and yet seemingly already 'there', potentially, in our cultural-theoretical resources. The challenge for present-day interpreters is to invent *dispositifs* of their own, which would articulate the elements and techniques that are so often abstracted into 'law' into assemblages that are as insightful and productive as 'imprisonment' or 'sexuality'. And who is to say that, even if we imagine theory as a kind of diplomatic initiative, lawyers themselves might not be more engaged by these yet-to-be-invented *dispositifs* than by some reconstitution of 'law' as rhetorical *techne*?

\* \* \*

Latour might have concluded from his ethnography of law making that the Conseil was the wrong place to begin in developing an actor-network cartography of law; or, alternatively, that there is less to law than social theories of law, or law itself, would suggest. Instead, he takes up the lawyer's sense of law making as a process of connecting enunciations and turns it into the paradigm of a novel dimension of sociality. So an almost pre-theoretical commitment to law trumps profound differences in theoretical architecture or strategy. In questioning this commitment, my point is not that we should go 'back to Foucault' in approaching the question of law, science, and 'material worlds'. Rather, it is that Foucault develops the only intensive treatment of law that does not presuppose 'law' as a basic social category. And, of themselves, in their very making or materiality, Foucault's *dispositifs* expose the contingency of that presupposition. A properly Foucauldian reflection on law and 'material worlds' might well unfold by way of a rich genealogy of 'law' as a social-scientific category, leading into an exploration of its retrenchment in the materiality of institutional and governmental *dispositifs*. My suggestion in making this more modest contrast between two takes on law and materiality is simply that we should begin with materiality rather than 'law'; and, in so doing, we should recognize that the vicissitudes of 'materiality' dissolve the instances – in this case 'law' – that they are supposed to constitute.